Integrated Pest Management

Integrated Pest Management

Edited by

J. Lawrence Apple
North Carolina State University at Raleigh
and
Ray F. Smith
University of California at Berkeley

PLENUM PRESS · NEW YORK AND LONDON

Library of Congress Cataloging in Publication Data

Main entry under title:

Integrated pest management.

"Most of the manuscripts in this volume were developed from papers presented in a symposium at the annual meeting of the American Association for the Advancement of Science held in San Francisco in February 1974."

Includes bibliographies and index.

1. Pest control—Congresses. I. Apple, Jay Lawrence, 1926- II. Smith, Ray Fred, 1919- III. American Association for the Advancement of Science.

SB950. A1157 632'.9 76-17549

ISBN 0-306-30929-7

A Division of Plenum Publishing Corporation
227 West 17th Street, New York, N.Y. 10011

Printed in the United States of America

Contributors

J. Lawrence Apple, *Departments of Plant Pathology and Genetics, North Carolina State University, Raleigh, North Carolina*

Dale G. Bottrell, *Department of Entomology, Texas A & M University, College Station, Texas; Presently: private entomological consultant Santa Rosa, California*

Philip S. Corbet, *Department of Biology, University of Waterloo, Waterloo, Ontario, Canada; Present Address: Department of Zoology, University of Canterbury, Christchurch, New Zealand*

Ellis B. Cowling, *Departments of Plant Pathology and Forest Resources, North Carolina State University, Raleigh, North Carolina*

D. W. DeMichele, *Department of Industrial Engineering, Texas A & M University, College Station, Texas*

H. C. Ellis, *Department of Entomology, North Carolina State University, Raleigh, North Carolina; Present Address: Georgia Coastal Plain Experiment Station, Tifton, Georgia*

William R. Furtick, *Plant Protection Service, The Food and Agriculture Organization of the United Nations, Rome, Italy*

J. D. Gilpatrick, *New York Agricultural Experiment Station, Geneva, New York*

Edward H. Glass, *Department of Entomology, New York State Agricultural Experiment Station, Cornell University, Geneva, New York*

S. C. Hoyt, *Tree Fruit Research Center, Washington Agricultural Experiment Station, Wenatchee, Washington*

Richard B. Norgaard, *Department of Agricultural Economics, University of California, Berkeley, California*

R. L. Rabb, *Department of Entomology, North Carolina State University, Raleigh, North Carolina*

Ray F. Smith, *Department of Entomological Sciences, University of California, Berkeley, California*

F. A. Todd, *Department of Plant Pathology, North Carolina State University, Raleigh, North Carolina*

William E. Waters, *Pacific Southwest Forest and Range Experiment Station, United States Forest Service, Berkeley, California; Present address: College of Natural Resources, University of California, Berkeley, California.*

Stephen Wilhelm, *Department of Plant Pathology, University of California, Berkeley, California*

Preface

The past decade is probably unparalleled as a period of dynamic changes in the crop protection sciences—entomology, plant pathology, and weed science. These changes have been stimulated by the broad-based concern for a quality environment, by the hazard of intensified pest damage to our food and fiber production systems, by the inadequacies and spiraling costs of conventional crop protection programs, by the toxicological hazards of unwise pesticide usage, and by the negative interactions of independent and often narrowly based crop protection practices. During this period, the return to ecological approaches in crop protection was widely accepted, first within entomology and ultimately within the other crop protection and related disciplines. Integrated pest management is fast becoming accepted as the rubric describing a crop protection system that integrates methodologies across all crop protection disciplines in a fashion that is compatible with the crop production system.

Much has been written and spoken about "integrated control" and "pest management," but to date no treatise has been devoted to the concept of "integrated pest management" in the broadened context as described above. Most of the manuscripts in this volume were developed from papers presented in a symposium at the annual meeting of the American Association for the Advancement of Science held in San Francisco in February, 1974. In arranging that symposium, the editors involved plant pathologists, entomologists, and weed scientists. This collection of manuscripts does not represent the "final word" on integrated pest management, but it should contribute to a common understanding of the concept as it applies to all crop protection disciplines.

J.L.A.
R.F.S.

Contents

Chapter VII

Chapter VIII

Chapter IX

I

The Origins of Integrated Pest Management Concepts for Agricultural Crops

Ray F. Smith, J. Lawrence Apple, and Dale G. Bottrell

Most discussions of the conceptual origins of integrated pest management (IPM) for crop protection center on the overuse and overdependence of chemical pesticides following World War II and their subsequent unfavorable consequences. Included among examples of these unfavorable consequences are the development of chemical-pesticide-resistant insect and plant pathogen populations, rapid resurgence of target pest populations following treatment, outbreaks of unleashed secondary pests, and undesirable environmental effects. Then as the story goes, this series of mishaps was countered by the wisdom of a few omniscient soothsayers in the form of pest management. Another account described it as a mixture of ''idealism, evangelism, pursuit of fashion, fund-raising, and even empire-building. The movement has indeed acquired the impetus and character of a religious revival. . . .'' (Price Jones, 1970).

There are elements of veracity in both of these versions of the modern origins of IPM, but the fundamental origins are less simplistic and more remote in history. The evolution of the concept and its terminology spans a period of several decades and has been influenced greatly by changing technologies and societal values. Some crop protection specialists continue to discredit the concept as representing only new jargon applied to long-established crop protection

---

RAY F. SMITH · Department of Entomological Sciences, University of California at Berkeley. J. LAWRENCE APPLE · Departments of Plant Pathology and Genetics, North Carolina State University at Raleigh. DALE G. BOTTRELL · Department of Entomology, Texas A & M University, College Station, Texas. Presently, private entomological consultant, 542 Riezzi Road, Santa Rosa, California 95401.

practices. We acknowledge that IPM is not a disjunct development in crop protection—it is an evolutionary stage in pest control strategy—but it represents a new conceptual approach that sets crop protection in a new context within a crop production system. Many components of the IPM concept were developed in the late nineteenth and early twentieth centuries, but IPM as now conceived is unique because it is based on ecological principles and integrates multidisciplinary methodologies in developing agroecosystem management strategies that are practical, effective, economical, and protective of both public health and the environment. The early efforts of crop protectionists to control pests with ecologically based cultural methods were not satisfactory; consequently, entomologists, plant pathologists, and later weed scientists were preoccupied with the discovery of pesticides that were economical and effective. Unfortunately, chemical methods were often not used to supplement cultural methods but to supplant them. Our state of technology and understanding of host–pest interactions has evolved to the point that an integration of pest control tactics for the control of a given class of pest (e.g., insects, plant pathogens, etc.) and for multiple classes of pests is not only feasible but necessary given the inadequacies of single-method, single-discipline approaches and their potential for undesirable effects on nontarget beneficial and pest species.

The concept of integrated control was first articulated by entomologists (Smith and Allen, 1954; Stern *et al.,* 1959) as an approach that applied ecological principles in utilizing biological and chemical control methods against insect pests. It was subsequently broadened to include all control methods (Smith and Reynolds, 1965). The idea of "managing" insect pests populations was proposed by Geier and Clark (1961), and "pest management" was advocated by Geier (1970) in preference to "integrated control." The concept of pest management has now been broadened to include all classes of pests (pathogens, insects, nematodes, weeds) and in this context is commonly referred to as IPM with the implication of both methodological and disciplinary integration.

Although IPM terminology evolved principally within the ranks of entomology, elements of IPM are deeply rooted in plant pathology. Even prior to the demonstration of the pathogenic nature of plant diseases in the nineteenth century, but principally after that time, methods for "managing" plant diseases were developed.

Evolution of Pest Control Practices

The history of man is a history of attempts to gain increasing control over the environment. At first this control was subminimal to the degree that poor shelter and unstable food supplies imposed population constraints. The gradual gain of man's capacity to control his environment parallels the gradual rise of civilization. But as man aggregated into villages and as he planted crops in clusters near rivers, he encountered increasingly severe attacks by pests against himself and his crops. For thousands of years, man could do nothing about these pests but appeal to the power of magic and a variety of gods. For the most part, he had to live with and tolerate the ravages of plant diseases and insects, but early man learned how to improve his conditions through "trial and error" experiences.

Prior to the emergence of the crop protection sciences and even before the biology of insects and the causal nature of plant diseases were understood, man evolved many cultural and physical control practices for protection of his crops. Many of these practices were subsequently proven to be scientifically valid even though they were derived mostly through empirical methods. Such methods now include sanitation (destruction or utilization of crop refuse, roguing of diseased plants, etc.), tillage to destroy overwintering insects and inoculum, removal of alternate hosts of pathogens and insects, rotation of crops to discourage buildup of damaging populations of insects and pathogens, timing of planting dates to avoid high-damage prone periods, use of insect- and pathogen-free seed and seedling methods, use of trap crops, selection of planting sites, pruning and defoliation, isolation from other crops, and management of water and fertilizers. Interestingly, laws were passed in Connecticut and Massachusetts to eradicate the barberry (Parris, 1968) as an alternate host for the stem rust of wheat (*Puccinia graminis*) even before this relationship was proven by DeBary in 1865.

The appropriate use of these cultural methods can reduce the damage potential to crops of essentially all pests and can provide economic control of many insect and disease pests. But there were many pests of high damage potential that could not be controlled adequately by early agriculturists by any combination of known cultural control methods. As biological knowledge grew during the eighteenth and nineteenth centuries and as pest problems became more severe resulting from an intensification of agriculture and the introduction of various pests into new areas, man became increasingly preoccupied with the search for more effective pest control measures. For plant diseases, this effort was stimulated especially by the introduction about 1872 of downy mildew (*Plasmopara viticola*) and powdery mildew (*Uncinula necator*) of grape from the United States into France on phylloxera-resistant root stocks. The result was devastating to the French wine industry. The late blight disease (*Phytophthora infestans*) epidemic on potatoes in Western Europe, especially in Ireland, during the 1840s resulted in widespread famine, starvation, and migration of human populations (Large, 1940). The need for effective control measures for diseases was urgent. The discovery during the 1850s that fungi may produce diseases of plants opened the way for the scientific study of agents to control diseases, and the principal search was for chemical compounds (Parris, 1968).

Various chemicals and concoctions were recommended for the control of insects and diseases as early as the eighteenth century (Lodeman, 1903). A liquid suspension of slacked lime and powdered tobacco was recommended during this period for the control of plant lice; various other concoctions of plant materials, animal manures, soot, dry ashes, sea water, urine, soap, turpentine, alcohol, farmyard drainage, and much more were recommended for insect and/or disease control. It appears that the gardeners of that day were of the persuasion that the effectiveness of a chemical or mixture was directly proportional to the offensiveness of its odor or taste.

No important changes took place in the materials used for the destruction of fungi or insects until about 1882 with Millardet's serendipitous discovery of "Bordeaux mixture" (quicklime and copper sulfate) in France. This set the stage for practical chemical control of plant diseases. Probably the destruction of the potato crop (*Solanum tuberosum*) in the United States by the potato beetle (*Leptinotarsa decemlineata*) prompted the discovery of Paris green about 1870, which ushered in the era of chemical insecticides (Lodeman, 1903). Much progress was made in the technique of chemical control of both plant diseases and insects during the last quarter of the nineteenth century—so much in fact that Lodeman (1903) was prompted to state that

> The best is generally the most profitable commodity, and the poorest is the least so; and the grower of today has it in his power to produce the best. It rests entirely with him whether his apples shall be wormy or not, whether his trees shall retain their foliage or lose it from disease. There are few evils that affect his crops which he cannot control, in many cases almost absolutely. Only a few diseases remain which still refuse to submit to treatment, but the number is rapidly decreasing, and the time will come when these also will disclose some vulnerable point which will allow for their destruction. . . . Foremost among the operations by means of which cultivated plants are protected from their enemies is spraying.

There was indeed optimism as early as the turn of the twentieth century that both diseases and insects would ultimately be controlled by chemical pesticides.

Cultural control of plant diseases was stimulated by the work of Sorauer in Germany, who was one of the first to demonstrate the predisposition of a host to infection by environmental conditions. This knowledge of the plant disease process is essential to the development of effective cultural control procedures since these aim chiefly at altering the environment as it affects the crop and pathogen, the interaction of crop and pathogen, and their interactions over time (Stevens, 1960).

Plants resistant to insects and diseases were recognized in the nineteenth century, but the deliberate development of pest-resistant varieties was not possible until after the rediscovery of Mendel's laws of heredity in 1900. Following this breakthrough, the approach was quickly exploited for the control of important plant diseases of cereal and other crops, but was pursued less vigorously for insect control until recent years.

The success with chemical control and host resistance in controlling plant pests distracted from the importance of cultural control in many instances. This situation prompted Stevens (1960) to conclude that the simplicity and general effectiveness of the host resistance and chemical control approaches had drawn attention away from cultural control to the point that it enjoyed less popular understanding and support. The plant pest control literature of the 1900–1965 period demonstrates clearly a preoccupation with the development of better resistant varieties (principally against pathogens) and better chemicals, but comparatively little attention was given to pathogen or insect ecology and cultural controls.

Early Advocates of an Ecological Approach to Pest Control

As the agricultural experiment stations emerged in the United States in the late nineteenth century, entomologists and plant pathologists began to discover biological explanations for the earlier empirically developed pest control methodology which had been restricted largely to natural and cultural measures, sometimes augmented by minimal use of the earliest insecticides or fungicides. Partly by intuitive insight and partly because there was little choice, leading entomologists advocated an ecological approach to pest control. In the 1880s, Stephen A. Forbes, State Entomologist of Illinois and Professor of Zoology and Entomology at the University of Illinois, adopted the word "ecology" and insisted upon the broad application of ecological studies in dealing with insect problems of agricultural crops (Metcalf, 1930). A number of others concerned with crop protection also advocated this fundamental approach.

In spite of this position by leading entomologists, there was over the next half-century a gradual erosion of the understanding of the importance of ecology in controlling insect pests. There were, of course, exceptions to this, and from time to time a plea was made for the ecological approach. Charles W. Woodworth, Professor of Entomology at the University of California, advocated an ecologically based pest management approach throughout his long career (Smith, 1975). For example, in 1896 (Woodworth, 1896) he stated that everyone should have a clear idea of the controls available and how to apply them:

> But it is equally essential that he should fully understand when to apply and not to apply. . . . Money or time should only be invested in (pest control) when there is

> good prospect of an ample return. It is safe to say that, even in California, where this matter has been agitated for so many years, in only a very small fraction of the cases where injury might be prevented is the proper treatment made. . . . On the other hand, it may also be said that when treatment is made it is often of no effect, and a waste of time and money. Careful observations of the practices in this State in reference to treating insects and fungi makes it appear that fully half of what it now costs to treat our crops is wasted.

Professor Woodworth (1908) also discussed the need for carefully evaluating each mortality factor and investigating the interactions of the separate components in terminology that clearly showed his familiarity with what we now call the "ecosystem concept" and "density-dependent" mortality. He was the first entomologist to point out the important fact that percent parasitism of an insect pest is not a valid criterion for assessing the efficacy of a parasite.

There were other early advocates of an "ecological approach" to insect pest control. In 1926 Charles Townsend, influenced by his experience in Peru, stated that "environmental investigations furnish the only sure basis for work leading to the speedy discovery of proper measures for the control of insects, whether for the suppression of injurious forms or for the extension of beneficial ones." In 1945, before the impact of DDT and the organic pesticides, Michelbacher also stressed the importance of ecology in insect control.

There was apparently no strong parallel concern among plant pathologists about the application of ecological principles to the "management" of plant diseases. After "cause and effect" relationships between pathogens and disease symptoms were established, it was generally recognized that pathogen life cycles should be understood to reveal a "weak link" that might be exploited for control. But probably due to the fact that plant pathologists were trained as botanists, they were not as concerned about the interaction of pathogen populations and their total environment as were some entomologists on behalf of insect populations.

Early Pest Management for the Cotton Boll Weevil

An analysis of the development of insect control in cotton also reveals the early foundations of IPM. The boll weevil, a native of Mexico, entered the United States in the late 1890s. Gradually the pest spread from its point of entry in South Texas into other states, and by 1922 it was distributed throughout practically the entire Cotton Belt, from Texas to the East Coast (Gaines, 1957). Research designed to control or eradicate the pest was begun in 1891 by an entomologist at the A & M College of Texas (now Texas A & M University) and in 1894 by USDA entomologists (Gaines, 1957). The earlier workers recognized the boll weevil problem as an extremely complex and serious one. Regardless of their initial notions about dealing with the pest, they soon rejected eradication as a realistic goal and commenced to develop what gradually evolved into a highly sophisticated system of pest management. It is difficult to determine whose influence was dominant in developing this system. Howard (1896), incorporating the results of field studies by Schwarz and Townsend, undoubtedly influenced others into looking at multicomponent suppression techniques. He stressed cultural control, especially the early fall destruction of cotton plants, and recommended early planting and clean cultivation. He also encouraged trapping weevils late in the fall and overwintering weevils early in spring and destruction of volunteer plants. Malley (1901), Hunter (1902, 1903) and Hunter and Hinds (1904) and several others (Dunn, 1964) further expanded the multicomponent management approach.

Though the term ''pest management'' was not mentioned in any reference on boll weevil prior to about a decade ago, the early workers not only preached but also practiced this approach. In 1901, Malley stressed the utilization of cultural controls in fighting the weevil. He stated in his 1901 publication that

> Another difficulty in securing a general acceptance of this method * lies in the fact that there is a small percentage of immature bolls which might yet open, but which the stock eats. Again the scarcity of pickers sometimes results in the planters being far behind with their picking. This is the planter's misfortune and not the fault of the method suggested. Much depends on their *management* (our emphasis) along this line.

Although Hunter (1903) argued that Malley's (1901) suggestion for insecticidal control of boll weevil was futile, Malley nevertheless continued for some time to recommend control with arsenic and arsenate of lead. But Malley (1901) himself obviously was aware of the limitations in using insecticide:

> It must be plain from the discussions in the foregoing page that spraying should not be depended upon solely, but in conjunction with the cultural methods. Neither *system* (our emphasis) used alone will attain the greater efficiency. If neither one is to be depended upon alone, the cultural methods are far more economical and efficient, and are capable of more general application under a greater variety of conditions. There can be no question of the desirability and the advantage of spraying, but it should be secondary, and should be practiced in conjunction with the cultural system.

Hence, by 1901, a system approach to cotton insect management had been advocated, and many components of the weevil's life system and cotton agroecosystem as well were understood. By 1904, Hunter and Hinds had presented a fairly good conceptual model of the pest's life system and recognized many

* The method Malley referred to was the grazing of livestock in cotton fields after harvest. The animals consumed green squares and bolls, thereby not only destroying immature boll weevils inside these forms but also reducing the adult weevil's food supply.

of the interdependent environmental factors causing a seasonal change in population density. A highly complex and sophisticated system had been fully developed as early as 1920, at which time economic thresholds were determined as guidelines for beginning treatments with calcium arsenate.

The classical publication of Hunter and Coad (1923) serves as a testimony to the advanced state of knowledge of boll weevil management in the early 1920s. This publication synthesized much that was known at the time about factors involved in controlling the pest. Although calcium arsenate was advocated as a method of control (Howard, 1919), Hunter and Coad cautioned against unilateral dependence on this material:

> Although the success of poisoning the weevil under certain conditions has been proven beyond doubt, there is danger that farmers may depend too much upon it and neglect the cultural practices which are absolutely essential in any system of weevil control. . . . Poisoning is supplementary and depends for its success upon the other necessary steps. . . . The crop itself must actually be made by other expedients; poisoning is merely a device to protect it when it is made.

In summing up the rules to follow in using calcium arsenate against the weevil, Hunter and Coad wisely recommended:

1. Start poisoning when the weevils have punctured 10 to 15% of the squares.
2. Then stop poisoning until the weevils again become abundant.
3. Do not expect to eradicate the weevils.
4. Always leave an occasional portion of a cut (of cotton) unpoisoned for comparison with the adjoining poisoned tract. This will show how much you have increased your yield by poisoning.

At the same time the concept of pest management was evolving for the boll weevil, basic concepts were being molded for managing other cotton pests too. Whitcomb (1970), in a review of the history of integrated control of cotton insects in the United States, pointed out that "Comstock's (1879) thoroughness in his studies of predators and parasites attacking cotton insect pests would shame many modern workers."

Many of these early management strategies were highly efficient and were adopted by the more progressive cotton farmers of the day.

Shift to Dependence on Chemicals and to a Lesser Extent on Resistant Varieties

In spite of the occasional warnings about the hazards of unilateral approaches to pest control, crop protection in the United States since 1920 has gradually shifted toward dependence on chemical pesticides and on disease-resistant varieties but to a less extent on insect-resistant varieties. The signals from populations of pests resistant to chemicals (e.g., red scale resistant to

HCN on citrus and codling moth to lead arsenate on apples) were ignored. The pattern of developments with cotton insect control during the period 1920–1945 is a good example of what happened.

It is difficult to pinpoint the causes of the paralysis of applied cotton entomology which began in the 1920s and peaked just a few years ago. Applied entomology as a whole suffered the same fate at about the same point. Perhaps the cause was a social one, and the remedy lay in public policies that were beyond the grasp of the age. On the other hand, the attitudes and mistakes of the entomologists added more than any other single factor.

Ironically, the beginning coincided with the time that a number of cotton entomologists discovered practical methods for applying calcium arsenate. These entomologists dropped their ecologically based work on cultural control, biological control, and resistant varieties and began exhaustive research on dusting schedules, dosages, swath patterns, and nozzle orifices. This shift in research emphasis paid tremendous dividends too, because control with the insecticides was spectacular. The applied entomologist became obsessed by the immense power he commanded over nature with this potent weapon.

The early 1920s to middle 1940s were definitely dominated by applied entomologists who adhered to and practiced this chemical approach. There were a few dissenters (notably, Isely and Baerg, 1924; Baerg *et al.*, 1938), who warned that the inorganic insecticides should be applied only as necessary to supplement other controls. Baerg *et al.* (1938), Gaines (1942), Bishopp (1929), Fletcher (1929), Sherman (1930), and Ewing and Ivy (1943) recognized that application of the inorganics to control the boll weevil accentuated the infestation of cotton aphid, *Aphis gossypii* Glover, and bollworm, *Heliothis zea* Boddie. Nevertheless, the unilateral insecticide approach dominated.

Initial Impact of Organic Pesticides

The prevailing philosophy adhered to by applied entomologists during the second quarter of this century was given yet even greater opportunities for expression when the post-World War II organic insecticides were introduced in the late 1940s. Entomologists enthusiastically adopted into their control programs DDT and other organochlorines and later organophosphorus (OP) and carbamate materials.

Whitcomb (1970) and Newsom (1970) described the approach that a majority of the cotton entomologists in the post-War period advocated for the new organic materials. In general, entomologists recommended that farmers spray their cotton once weekly from the time it started squaring until near harvest, but there was no method by which the farmer could determine if the spray was actually needed. Certainly, some entomologists had more influence than others in promoting this method of control. Nevertheless, this approach was favorably accepted by most entomologists of the day. Articles by Rainwater (1952), Gaines (1952, 1957), Curl and White (1952), and Ewing (1952) expressed the general philosophy which prevailed during the first 5 to 15 years following introduction of the organochlorines.

Despite this prevailing philosophy, there also were some very sound insect management programs developed during this period. A classical example was the cultural control program developed for pink bollworm, *Pectinophora gossypiella* Saunders, recently reviewed by Adkisson (1972). Also, several entomologists continued to encourage farmers or scouts to check the cotton fields regularly and to apply insecticides only when the pest population reached economic thresholds (G. L. Smith, 1953; R. F. Smith, 1949; Boyer *et al.*, 1962). And Newsom and Smith (1949) and a few of their followers gathered evidence that the new organic insecticides seriously affected the populations of several natural enemies. In general, however, the period from the late 1940s to the early mid-1960s marked a time when most major cotton growing areas troubled with severe insect pests came under a heavy blanket of insecticide.

The rest of the story has been documented recently in publications by Stern (1969), Adkisson (1969, 1971, 1972, 1973*a,b*), Smith and van den Bosch (1967); van den Bosch *et al.* (1971), Smith (1969, 1970, 1971), Doutt and Smith (1971), and Newsom (1970). All of these articles pointed to the problems which eventually arose as a result of overuse of the insecticides introduced after World War II.

Return to Ecological Approaches in Pest Control

The pest problems of agricultural crops in the United States have been aggravated and intensified by a complex of factors. Some of the factors were related to the limited base of tactics employed (primarily chemicals and host resistance). In many instances these no longer controlled the target pests, interfered with the control of other pests, and released species from existing natural control so that they became pests. In some cases, the chemicals modified the physiology of the crop plants unfavorably, created hazards to man's health, destroyed pollinators and other desirable wildlife, and in other ways produced undesirable effects.

The introduction of new pest species into the U.S. agroecosystems has also placed stress on an already overburdened pest control technology. Furthermore, the pressure on agroecosystems toward greater intensity of production over the years has forced them to evolve very rapidly and has created new environments for the pests. As a result the agroecosystems often have become more vulnerable to pests. Changes in tillage, water management, crop varieties, fertilization, and other agronomic practices greatly influenced pest incidence, very often in favor of the abundance of the pest species. The increased complexity and intensity of agricultural production practices along with reduced genetic diversity in many agricultural crop species combined to produce a new magnitude of crop hazards.

The intensification and increasing complexity of crop protection problems coupled with the associated environmental, financial, and health hazards of heavy chemical usage have combined to stimulate great interest in the importance of crop protection and in the broad ecological approach as a sound approach to acceptable solutions. And also of great importance has been the increased financial support for pest management research, extension and field-implementation programs.

As the problems intensified in crop protection, the debate over the matter also intensified. This finally erupted in the *Silent Spring* episode (Carson, 1962; Westcott, 1962). The President's Science Advisory Committee issued a special report in 1963 entitled *Use of Pesticides* that found fault with a number of crop protection chemicals, especially insecticides. The Southern Corn Leaf Blight epidemic of 1970 (Tatum, 1971) emphasized the problem of the genetic vulnerability of major crop species in the United States to damaging attacks by pests (NAS, 1972). These situations combined with increased awareness of a world food crisis motivated government and institutional actions supportive of the development of IPM systems for major agroecosystems in the United States.

A major step toward development of IPM programs was taken by the Federal Government in 1972. In his Message on Environmental Protection, the President of the United States directed the cognizant agencies of government to take immediate action toward development of pest management programs in order to protect (a) the Nation's food supply against the ravages of pests, (b) the health of the population, (c) and the environment. The President's Directive prompted funding of a national research project involving 19 universities and various federal agencies entitled *The Principles, Strategies, and Tactics of Pest Population Regulation and Control in Major Crop Ecosystems* (known as the "Huffaker IPM Project"). Other programs initiated in 1972 were pilot projects for implementing pest management programs in the various states; curriculum development for training and certification of crop protection specialists by the land-grant universities; and pilot pest management research projects within the Agricultural Research Service in collaboration with state groups. These actions have been followed with an intensification of pest management research within state agricultural experiment stations and federal agencies financed by both state and federal sources.

The Modern Integrated Pest Management Approach

The great advances made in recent decades to provide more food and other agricultural products serve as testimony to man's ability to develop and manipulate scientific technology to his advantage. However, experience has proven repeatedly that technology often has been developed and manipulated to satisfy only short-term needs. The unfortunate consequences of two of these technological advances have provided much of the underlying rationale for IPM. One of these advances emerged shortly after World War II with the introduction of DDT and other pesticides. The other of these advances was the introduction and world-wide use of the high-yielding but narrow genetically based crop varieties that now serve as the basis of the "green revolution."

It hardly seems appropriate here to discuss in detail the consequences already experienced from these two technological advances or the implications of this technology for the future. The appearance of numerous insecticide-resistant strains of crop pests and the growing evidence of new biotypes or strains of pests that resist and adapt to previously pest-resistant crop varieties adequately portray the consequences and shortcomings of technological advances aimed at satisfying short-term needs. These unfortunate experiences have come about because technology was pursued unilaterally, indifferent to the potential countermanding ecological consequences that have already caught up with us in some instances. A benefit has evolved from these unfavorable consequences, however, in the form of closer attention by all crop protection specialists to those ecological principles that are basic to successful long-term pest control.

Scientific pest control has always required a knowledge of ecological principles, the biological intricacies of each pest, and the natural factors that tend to regulate their numbers. Today, it is more necessary than ever before to take a broad ecological overview concerning these problems, and to consider all possible factors, both natural and artificial, that can be used against crop pests. We cannot afford any longer to disregard the considerable capabilities of pest organisms for countering control efforts. Most of these organisms are extremely fitted to thrive in our agroecosystems and to adapt to changing crop production conditions. It is for this prudent reason that we must understand Nature's methods of regulating populations and maximize their application. This clearly requires a realignment of research objectives and practices. A unified, balanced approach is needed, predicated mainly on widely proven principles of pest control and ways of implementing them, and recognizing the limitations as well as the advantages of any new methods that are evolved.

Hence, the IPM activity is attempting to bring together teams of scientists who are capable of taking a broad ecological overview of the pest problems associated with our major agroecosystems and who are willing to assume leadership in developing unified, ecologically based approaches to pest control. These multidisciplinary teams will have the potential for developing effective, economical, and long-term solutions for agricultural pest problems if they approach cropping systems as ecological units and if they apply the most sophisticated methods of experimentation, synthesis, and analysis.

One of the many pitfalls of unilaterally designed pest control that focuses on only one pest or a closely related group of pests is the result of the countermanding effects of a similarly designed control to remedy the emergence of a different pest or pest complex. An extremely effective unilateral control strategy developed for a given crop insect pest, for example, may be totally negated on the arrival of a new pest inhabitant such as a weed, plant pathogen, spider mite, or other insect. For that matter, the arrival of a new pest in a given cropping district may even negate the effect of highly sophisticated IPM programs. A classic example exists in the desert cotton-producing areas of California and Arizona. Effective integrated control programs had been developed for the cotton insect pests native to these areas and were being successfully adopted by many of the area farmers until arrival of the pink bollworm in the late 1960s. The prophylactic action to control this invader, which consisted principally of wholesale insecticidal applications, totally disrupted and counteracted benefits of the previously adopted integrated control programs. This is one of several examples that demonstrate the need for built-in safeguards to prevent negation of an operational IPM strategy when confronted with an invasion by a new type of pest or the evolution of new biotypes of old pests. This will be no easy task since the arrival of these new situations cannot be readily predicted. However, the emerging IPM strategies offer superior safeguard mechanisms through surveillance systems, but these are conditioned by our limited understanding of highly complex, evolving pests, pest mobility, pest interactions, and shifting germ plasm of agricultural crops. First, the holistic approach has provided much new insight into the complexities of interspecific relationships of pests and potential pests in the agroecosystems. For example, the major natural enemies and other population-regulating factors of key pests, occasional pests, and potential pests of many major cropping systems have been identified. Hence, we have become more aware of the sensitivity of these natural interrelationships and have a much clearer understanding of the disruptive, catastrophic events (such as insecticides, plant varieties, cultural practices, etc.) that may upset this balance. Second, today's IPM programs are bringing together teams comprising specialists in crop protection and other supportive disciplines. This improved interdisciplinary communication and interaction should prevent the emergence of unilateral counteracting control measures as exemplified by the following example. Currently, in the southern United States, a fungicide (benomyl) is being applied to large acreages of soybeans to control a complex of plant pathogens without knowledge of the specific pathogens being controlled or the economic damage actually being caused by them. Unfortunately, this wholesale "tonic treatment" is negating the otherwise highly effective integrated control programs against soybean insects. For, in addition to control of the plant pathogenic fungi, the fungicide has adverse effects on populations of entomopathogenic fungi that are important natural enemies of several lepidopterous insect pests of soybean. More disrupting yet, however, is the inclusion of broad spectrum "insurance" insecticides with the fungicide treatments. Though this problem has yet to be resolved, entomologists working in concert with plant pathologists in the course of developing IPM programs are in a much better position than ever before to develop a satisfactory solution. Third, the IPM strategies under development promise to provide options and pathways to alternate solutions in the event of "crisis" developments such as the emergence of pesticide-resistant strains, biotypes that can overcome host resistance, or arrival of pest invaders in an ecosystem. The optional approaches will be developed, however, only through the continued close collaboration of scientists in many disciplines focusing on holistic systems. But they will be essential to safeguard against disrupting crises that are certain to emerge in pest management programs of the future because dynamic agroecosystems provoke counteracting changes in the dynamic pest subsystems that must be "managed" to preserve our crop production potential.

Literature Cited

Adkisson, P. L., 1969, How insects damage crops, *in* How Crops Grow a Century Later, *Conn. Agr. Exp. Sta. Bull.* **708:**155–164.

Adkisson, P. L., 1971, Objective uses of insecticides in agriculture, *in:* Proceeding of Symposium on Agricultural Chemistry—Harmony or Discord for Food, People and the Environment (J. E. Swift, ed.), University of California Division of Agricultural Science, pp. 110–120.

Adkisson, P. L., 1972, Use of cultural practices in insect pest management, *in:* Implementing Practical Pest Management Strategies—Proceedings of the National Extension Insect–Pest Management Workshop, Purdue University, March 14–16, 1972, pp. 37–50.

Adkisson, P. L., 1973*a*. The integrated control of the insect pests of cotton, *in:* Proceedings of the Tall Timbers Conference on Ecological Animal Control by Habitat Management, No. 4, pp. 175–188.

Adkisson, P. L., 1973*b*, The principles, strategies and tactics of pest control in cotton, *in: Insects: Studies in Population Management* (P. W. Geier, L. R. Clark, D. J. Anderson, and H. A. Nix, eds.), Ecological Society of Australia (Memoirs 1), Canberra. pp. 274–283.

Baerg, W. J., D. Isley, and Sanderson, M. W., 1938, *Ark. Agr. Exp. Sta. Bull.* **368:**62–66.

Bishopp, F. C., 1929, The bollworm or corn earworm as a cotton pest, *USDA Farmers' Bull.* **1595,** 14 p.

Boyer, W. P., Lincoln, C., and Warren, L. O., 1962, Cotton scouting in Arkansas, *Ark. Agr. Exp. Sta. Bull.* **656,** 40 p.

Carson, R., 1962, *Silent Spring*. Hamish Hamilton, London. 304 p.

Comstock, J. H., 1879, *Report upon Cotton Insects,* U.S. Government Printing Office, Washington, D.C. 511 p.

Curl, L. F., and White, R. W., 1952, The pink bollworm, *USDA Yearbook Agr.:* 505–511.

Doutt, R. L., and Smith, R. F., 1971, The pesticide syndrome, *in: Biological Control* (C. B. Huffaker, ed.) Plenum Press, New York, pp. 3–15.

Dunn, H. A., 1964, Cotton Boll Weevil (*Anthonomus grandis* Boh.): Abstracts of Research Publications, 1843–1960, U.S. Department of Agriculture Cooperative State Research Service, Miscellaneous Publication 985, 194 p.

Ewing, K. P., 1952, The bollworm, *USDA Yearbook Agr.:*511–514.

Ewing, K. P., and Ivy, E. E., 1943, Some factors influencing bollworm populations and damage, *J. Econ. Entomol.* **36:**602–606.

Fletcher, R. K., 1929, The uneven distribution of *Heliothis obsoleta* (Fabricius) on cotton in Texas, *J. Econ. Entomol.* **22:**757–760.

Gaines, R. C., 1942, Effect of boll weevil control and cotton aphid control on yield, *J. Econ. Entomol.* **35:**493–495.

Gaines, R. C., 1952, The boll weevil, *USDA Yearbook Agr.:* 501–504.

Gaines, R. C., 1957, Cotton insects and their control, *Ann. Rev. Entomol.* **2:**319–338.

Geier, P. W., 1970, Organizing large-scale projects in pest management, *in:* Meeting on Cotton Pests, Panel of Experts on Pest Control, FAO, Rome, September, 1970. 8 p.

Geier, P. W., and Clark, L. R., 1961, An ecological approach to pest control, *in: Proceedings of the Eighth Technical Meeting, International Union for Conservation of Nature and Natural Resources,* Warsaw, 1960, pp. 10–18.

Howard, L. O., 1896, The Mexican cotton boll weevil, U.S. Department of Agriculture Bureau of Entomology, Circular No. 14. 8 p.

Howard, L. O., 1919, Report of the entomologist, U.S. Department of Agriculture Bureau of Entomology, Report No. 27. 27 p.

Hunter, W. D., 1902, The present status of the Mexican cotton boll weevil in the United States, *USDA Yearbook Agr.:* 369–380.

Hunter, W. D., 1903, Methods of controlling the boll weevil, *USDA Farmers' Bull.:***163,** 16 p.

Hunter, W. D., and Coad, B. R., 1923, The boll weevil problem, *USDA Farmers' Bull.:* **1329,** 30 p.

Hunter, W. D., and Hinds, W. E., 1904, The Mexican cotton boll weevil, *USDA Div. Entomol. Bull.:* **45,** 116 p.

Isley, D., and Baerg, W. J., 1924, The boll weevil problem in Arkansas, *Ark. Agr. Exp. Stn. Bull.:* **190,** 22 p.

Large, E. C., 1940, *The Advance of the Fungi,* Holt, New York. 488 p.

Lodeman, E. G., 1903, *The Spraying of Plants,* MacMillan, New York. 399 p.

Malley, F. W., 1901, The Mexican cotton-boll weevil, *USDA Farmers' Bull.* **130.** 29 p.

Michelbacher, A. E., 1945, The importance of ecology in insect control, *J. Econ. Entomol.* **38:**129–130.

Metcalf, C. L., 1930, Obituary, Stephen Alfred Forbes, May 29, 1844–March 13, 1930, *Entomol. News* **41(5):**175–178.

NAS, 1972, Genetic vulnerability of major crops, Committee on Genetic Vulnerability of Major Crops, National Academy of Science, Washington, D.C. 307 p.

Newsom, L. D., 1970, The end of an era and future prospects for insect control, *in:* Proceedings of the Tall Timbers Conference on Ecological Animal Control by Habitat Management, No. 2. pp. 117–136.

Newsom, L. D., and Smith, C. E., 1949, Destruction of certain insect predators by applications of insecticides to control cotton pests, *J. Econ. Entomol.* **42:**904–908.

Parris, G. K., 1968, *Chronology of Plant Pathology,* Johnson & Sons, Starkville, Miss. 167 p.

Price Jones, D., 1970, Applied biology as an evolutionary process, *Ann. Appl. Biol.* **66:**179–191.

Rainwater, C. F., 1952, Progress in research on cotton insects, *USDA Yearbook Agr.*:497–500.

Sherman, F., 1930, Results of airplane dusting in the control of cotton bollworm (*Heliothis obsoleta* Fab.), *J. Econ. Etnomol.* **23:**810–813.

Smith, G. L., 1953, Supervised control of cotton insects, *in:* Conference on Supervised Control of Insects, Department of Entomology and Parasitology, University of California, Berkeley, February 6–7, 1953.

Smith, R. F., 1949, Manual of supervised control, Division of Entomology and Parasitology, University of California, Berkeley. 27 p.

Smith, R. F., 1969, The new and the old in pest control, *Proc. Accad. Nazion. dei Lincei, Rome (1968)* **366(138):**21–30.

Smith, R. F., 1970, Pesticides: their use and limitations in pest management, *in: Concepts of Pest Management* (R. L. Rabb and F. E. Guthrie, eds.), North Carolina State University, Raleigh, p. 103–11.

Smith, R. F., 1971, Economics of pest control, *in:* Proceedings of the Tall Timbers Conference on Ecological Animal Control by Habitat Management, No. 3, p. 53–83.

Smith, R. F., 1975, The origin of integrated control in California—an account of the contributions of C. W. Woodworth, *Pan Pacific Entomol.* **50(4):**426–429.

Smith, R. F., and Allen, W. W., 1954, Insect Control and the Balance of Nature, *Sci. Am.* **190(6):**38–42.

Smith, R. F., and Reynolds, H. T., 1965, Principles, definitions and scope of integrated pest control, *in:* Proceedings of the FAO Symposium on Integrated Pest Control, Vol. 1, pp. 11–17.

Smith, R. F., and van den Bosch, R., 1967, Integrated control, *in: Pest Control: Biological, Physical, and Selected Chemical Methods* (W. W. Kilgore and R. L. Doutt, eds.) Academic Press, New York, p. 295–340.

Stern, V. M., Smith, R. F., van den Bosch, R., and Hagen, K. S., 1959, The integrated control concept, *Hilgardia* **29:**81–101.

Stern, V. M., 1969, Interplanting alfalfa in cotton to control lygus bugs and other insect pests, *in:* Proceedings of the Tall Timbers Conference on Ecological Animal Control by Habitat Management, No. 1, p. 55–69.

Stevens, R. B., 1960, Cultural practices in disease control, *in:* Plant Pathology: An Advanced Treatise (J. G. Horsfall and A. E. Dimond, eds.), Academic Press, New York, Vol. 3, pp. 357–429.

Tatum, L. A., 1971, The southern leaf blight epidemic, *Science* **171:**1113–1116.

van den Bosch, R., Falcon, L. A., Gonzales, D., Hagen, K. S., Leigh, T. F., and Stern, V. M., 1971, The developing program of integrated control of cotton pests in California, *in: Biological Control* (C. B. Huffaker, ed.), Plenum Press, New York, pp. 377–394.

Wescott, C., 1962, Halftruths or whole story, a review of *Silent Spring,* Manufacturing Chemist's Association, Washington, D.C.

Whitcomb, W. H., 1970, History of integrated control as practiced in the cotton fields of the south central United States, *in:* Proceedings of the Tall Timbers Conference on Ecological Animal Control by Habitat Management, No. 2, pp. 147–155.

Woodworth, C. W., 1896, Remedies for insects and fungi, *Calif. Agr. Exp. Sta. Bull.* **115:**1–16.

Woodworth, C. W., 1908, The theory of the parastitic control of insect pests, *Science* **28(712):**227–230.

II

Integrating Economics and Pest Management

Richard B. Norgaard

The pesticide problem is a social problem. The social objectives of nourishment, health, and environmental quality can be attained more efficiently by the implementation of integrated pest management techniques than through current crop protection practices. Solutions to the problem will come through changing the behavior of men. Farmers will selectively reduce pesticide use, select pesticides with a more narrow action spectrum, and utilize new biological and other control methods. The basic behavior of insects and weeds (ignoring adaptation such as resistance) and of chemicals will remain unchanged. The objective of this paper is to introduce plant protection scientists to economic principles and problems related to the design and implementation of optimal pest management strategies that will also reduce the pesticide problem.

While biological scientists are increasing their effort to understand the impact of pesticides in the agroecosystem, economists are just beginning to consider seriously the implications of pest management strategies to the farmer, the agricultural sector, and society. Largely due to the nationwide integrated pest management research project (IPM Project) initiated and directed by Carl Huffaker and Ray Smith of the University of California, the number of economists involved in pest management research in the United States has recently increased from less than a dozen to perhaps twice that number. However, few economists are working full-time in the area of pest management, and our efforts are dispersed geographically and by crop.

The overall objective of the IPM Project is to develop integrated pest management strategies—i.e., optimum combinations and use of all known inputs

---

RICHARD B. NORGAARD · Department of Agricultural Economics, University of California at Berkeley 94720. This is Giannini Foundation Paper No. 414.

and techniques including biological controls, cultural practices, and chemical approaches. Economics may enter into the design of these strategies in three interrelated ways: (1) the pest management goals of farmers are largely economic; (2) as a science of resource allocation, economics can aid in selecting optimal quantities and combinations of pest management inputs; and (3) the economist's understanding of the incentives underlying farmers' behavior and the effect on these incentives of alternative social institutions can speed the adoption of new pest management practices. The first role of economics is well recognized in the area of plant protection. The well-established subdiscipline of economic entomology and more recently the concept of the economic threshold incorporate this aspect of economics in their definitions. Entomologists have encouraged economists to enter the field of pest management with the expectation that economists would compare the benefits and costs of a few alternative strategies designed by biological scientists and declare one better than the others. The economists, however, have been more interested in the science of economics, of optimization and economic behavior, than in accounting.

Optimization implies an objective and a degree of control over pest management variables. Pest management objectives and the variables over which the decision-maker has some control are different, however, for the farmer, for institutions within the agricultural sector, and for society. For better or worse, decisions made at each level affect the range of choice at other levels. Both the sciences of economics and of ecology recognize and attempt to deal with the fact that everything is connected to everything else. Consequently, economists cannot simply discuss pest management on the farm without reflecting on all levels of decision making. Media and temporal constraints dictate, however, that this paper must proceed sequentially from a beginning to an end. The objectives of farmers and the pest management variables over which they can exercise some control are presented in the beginning, and the variables over which existing or potential pest management institutions within the agricultural sector might choose to exercise control are considered in the end. Broader consumer, farm worker, and environmental goals and policy variables, though very important, are not addressed in this paper.

Farm Strategy

A basic goal of commercial farmers is to maximize farm profits, the difference between returns and costs. The returns to the individual farmer from pest management are the increase in the money value of the yield at harvest resulting from a particular pest management strategy. By expressing the yield in monetary terms, both quality and quantity differences in the yield are simultaneously accounted for, and the change can be compared with costs. Costs of pest management which the individual farmer can affect include, for example, (1) the costs of acquiring information and making decisions, (2) the costs of carrying out cultural practices or applying inputs, and (3) the costs of pest control inputs. Expressed mathematically, the profit function is simply

$$\pi = pF(x) - C(x) \tag{1a}$$

where π represents the profits, p is the price of the product, $F(x)$ is the quantity of product expressed as a function of x, x is a vector of pest management inputs, and $C(x)$ is the cost of pest management inputs expressed as a function of x.

Profits are maximized when the derivative of π with respect to x is set equal to zero.

$$\frac{d\pi}{dx} = p\frac{dF(x)}{dx} - \frac{dC(x)}{dx} = 0 \tag{1b}$$

At this level, the incremental return from a small increase in pest management intensity is equal to the incremental cost (Figure 1). Incremental costs rise with increases in pest management intensity because of the higher costs of gathering more precise information, applying inputs more selectively or completely, or using more specialized materials. The incremental benefits decrease with increased pest management intensity as the remaining portion of the crop to be saved decreases. Profits are maximized where these curves intersect at level X_0. The profits to the farmer are represented by the difference between these two curves, the area represented by the triangle *abc*. Clearly, as long as the incremental costs of pest management are positive to the left of the intercept of the incremental benefit curve with the X axis at X_2, a profit-maximizing farmer should choose to lose some of his crop to pests rather than forego the even higher costs of stricter pest management.

The level of product and input prices affects the optimum intensity of pest management. For example, when farm prices rise, the returns to pest management rise, including the incremental returns from additional action. This is also illustrated in Figure 1 by the higher, new incremental return curve. More intensive pest management is profitable since the incremental returns intersect incremental costs at a higher intensity level, X_1. Similarly, an increase or decrease in pest management costs would decrease or increase the optimum level of management intensity, respectively.

It is a very difficult task to integrate even these most simple of economic concepts with a most simple biological model to determine the optimum timing and application of a single input such as a pesticide. A major difficulty lies in the fact that even the simplest of biological systems is more complex than most human technological production systems. Two years ago Headley (1972) re-

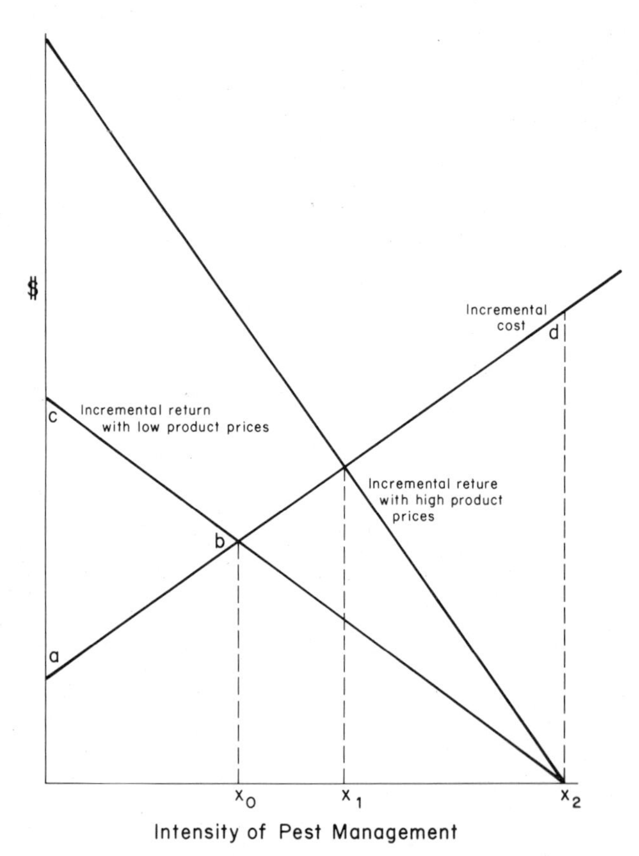

Figure 1. Optimal pest management. Profits are maximized when the difference between returns from pest management (the area beneath the incremental return line) and the costs of pest management (the area beneath the incremental cost line) is maximized. This occurs at the intensity of management where the incremental return and incremental cost are equal.

fined the definition of the economic threshold. Hall and Norgaard (1973) and Hueth and Regev (1974) have made further improvements with the use of simple functions for pest population growth, pest damage, pest control, and control cost. Since pesticide applications are almost always discrete rather than continuous, the concept of the threshold immediately splits into two components: (1) at what pest population level should control be initiated; and (2) by how much should the pest population be reduced? This is illustrated in Figure 2 by the decision to spray at t_1 and reduce the population from $P_{t_{1-\Delta}}$ to P_{t_1}. Economic entomologists have been concerned with the first aspect of the threshold but have largely ignored the second, the appropriate level of kill, on the assumption that the highest "practical" kill is optimal.

The relationship between the aggregate concept of pest management intensity and these two components of the threshold is complex. More intensive pest management means an increase in expenditures or effort. If pesticides were the only input, then more pesticides would be used. The threshold would, therefore, be "reduced." With respect to the two components of the threshold, spraying would be initiated at lower pest population levels, and the population would be reduced to a lower level than before. These conclusions, however, are based on the simple economic and biological models which are only now being explored.

The relationship between the damage threshold—the pest population level above which economic loss occurs— and the economic threshold deserves clarification. If controls are successfully initiated at the damage threshold, zero damage will occur. This is represented by management intensity level X_2 in Figure 1. Compared to the optimal level X_0, the farmer would take a loss represented by the triangle X_2bd because the costs of eliminating all pest damage are greater than the returns. Control costs were included in the definition of

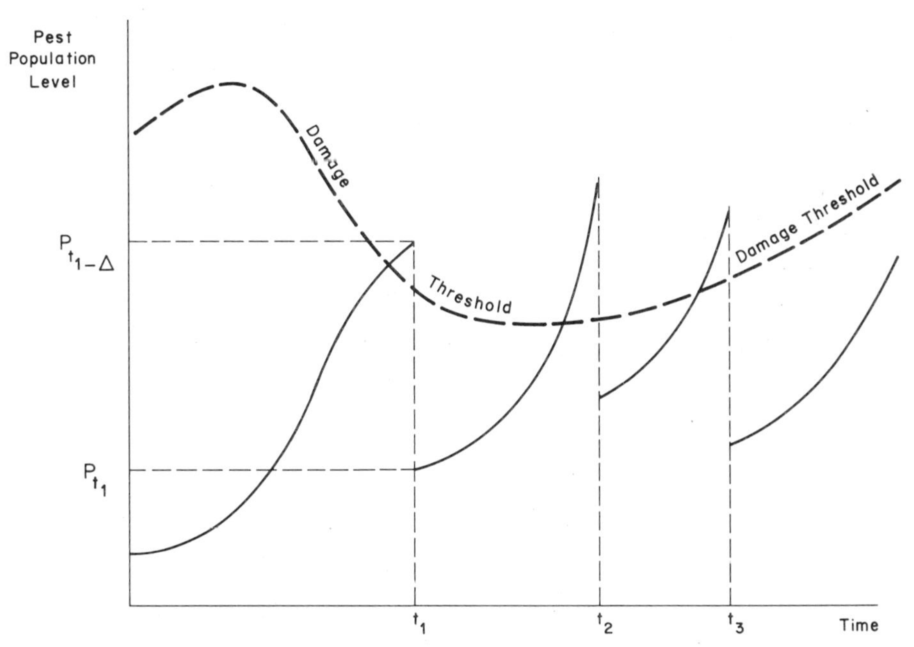

Figure 2. Optimal pest management. The peak infestation level should exceed the damage threshold when there are postive incremental costs of pest control.

the economic threshold, at least as early as 1959 in an article by Stern *et al.* A hypothetical damage threshold has been included in Figure 2 (Davidson and Norgaard, 1973). It should be noted that pest levels have been allowed to exceed this threshold, which indicates that the strategy chosen recognizes the trade-off between damage and control costs. Control programs with costs that can be varied through changing the number of applications of the control input or the concentration of the input per application should allow some damage to occur simply because the incremental costs of the program are positive. Pest management strategies whose costs are fixed—either they are implemented or not implemented—may not ever have pest population levels above the damage threshold. Many cultural practices to control pests fit into this category. For these strategies, the term economic threshold has little meaning.

The concept of the economic threshold becomes even more complex when we realize that the damage threshold, integral to the economic threshold concept, can itself be varied by crop management utilizing economic inputs. Indeed, major advances in "pest management," such as altered planting dates, improved timing of irrigation and cultural practices, and resistant varieties, affect the economic threshold through the damage threshold. It is becoming increasingly difficult to adapt the simple concept developed by Stern *et al.* (1959) to the problem of optimizing truly integrated pest-crop management strategies. We now recognize that the threshold depends on numerous variables, that it is no longer simply a pests-per-sweep concept. The term "threshold," however, by its very nature and history, still connotes the time for a pest control action. But the strategies under development will likely interact with nonpest management objectives such as the rate of plant development. Hence, pest management decisions are likely to become less and less separable from crop management decisions in general. This development is inherent in the rising recognition of the "agroecosystem" concept. The concept of the economic threshold has played a very important role in limiting the overuse of pesticides but will probably decline in the future as integrated control techniques become truly integrated with crop management.

Though the long-run goal of farmers is to maximize profits, the costs of not meeting or having to renegotiate year-to-year contracts with marketing firms and banks and the discomfort of having to reduce personal expenditures make farmers risk averse. An uneven stream of returns over a 10-year period may be valued less than an even stream of equal amount because, in low-return years, the farmer must bear the costs of renegotiating with a lender, "getting by" with less capital to run his farm, and a lower personal income, and perhaps losing his farm. The avoidance of risk is clearly a criterion that must be weighed against profit maximization over the long run. Friedman and Savage (1948) originally illustrated this trade-off between certan and irregular monetary income as shown in Figure 3. The curve $U(I)$ expresses a relationship be-

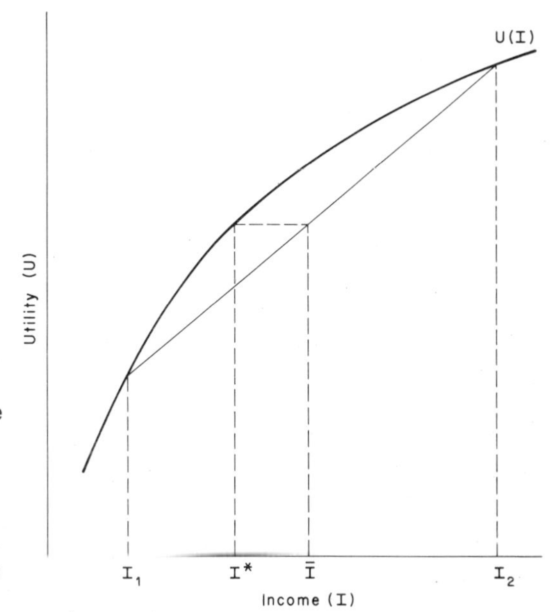

Figure 3. General relationship between utility and income. An uncertain income (0.5 probability of I_1 and 0.5 probability of I_2) with an expected value of $\bar{I}$. The utility of the uncertain income is equivalent to the certain income I^*, which is less than $\bar{I}$.

tween monetary income and utility or happiness. The decreasing slope of this curve indicates risk averseness. Faced with an even bet of receiving the high income of I_2, or the low of I_1, the farmer would be equally happy with the certain income of I^*, which is less than the average of I_1 and I_2. The difference between $\bar{I}$ and I^* indicates the premium the farmer would be willing to pay to avoid the uncertainty.

It seems to be a widely held view (Carlson, 1970; Hillebrandt, 1960; Smith, 1971; van den Bosch, 1972) that a substantial portion of total pesticide applications occurs for insurance purposes and that perceived risk and risk aversity are the major determinants of whether or not farmers adopt new pest management strategies. This view quite naturally leads to proposing pest damage insurance schemes as a substitute for pesticides to reduce use and to speed the adoption of new techniques. The relative importance of risk averseness has been tested in an economic analysis of the adoption of pest management consultants by California farmers (Willey, 1974). The real and perceived risk of following the advice of a chemical seller *versus* a pest management consultant and the farmer's risk averseness were measured through elaborate interviews with 160 farmers. Farmers who utilized consultants perceived greater risks from pest damage and were also more averse to risk than farmers utilizing sellers. Their perception of and averseness to risk may lead them to seek the information and advice provided by consultants to help reduce their risk. Risk is further reduced since the consultant's strategy is an additional option which does not preclude switching to a pesticide strategy at any time during the season. Conversely, the farmer who initiates the season with a pesticide strategy bears greater risk since he has less information and no option to change strategies during the season.

Much of the current research being conducted by entomologists and systems analysts is likely to reduce risk. Considerable real and perceived risk is now borne or countered by the farmer because of the high cost of determining insect population levels and the uncertain relationship between present population levels and future pest damage. The determination of the relationships between pheromone trap counts and population levels in the field and the construction of pest management simulation models are excellent examples of this type of research. As the costs of gathering information and making correct decisions decline over time, knowledge should be substituted for pesticides as insurance against risk.

Regional Strategies

Unlike most farm production problems, there are important relationships between farmers. The insects and management practices of one farmer can beneficially or detrimentally affect the insects and the success of the management practices of other farmers. Expressed mathematically, the profit functions of two farmers take the form:

$$\pi_A = pF_A(X_A, X_B) - C(X_A)$$
$$\pi_B = pF_B(X_A, X_B) - C(X_B)$$

where π_A and π_B are the profits of farmers A and B, respectively, and X_A and X_B are the pest management inputs of each farmer.

Since farmer A can only manage pests in his own fields, he can only maximize profits with respect to his pests and control inputs. Yet, farmer A's profits are a function of farmer B's pests and control inputs as well. The situation is shown as symmetrical for farmer B though the functional relationship may be quite different if farmer B is raising a different crop. Acting individually, these farmers will equate their own incremental costs with their own incremental returns. If each farmer would also consider the benefits or costs imposed by his pests and management practices upon the other farmer, their collective profits can be greater. But farmers acting as individuals have no incentive to behave this way. Some form of organization between farmers is required to capture these gains.

Examples of these interrelationships are numerous. Each farmer's application of pesticides decreases the pool of susceptible insects; but the effects of increased resistance, resurgence, and insecticide-induced secondary pest outbreaks are felt by all farmers who have to contend with the pest complex. In California we are learning that, if lygus are controlled in safflower and alfalfa fields before they move into cotton fields, then secondary outbreak of thrips from spraying lygus on cotton can be avoided. This strategy, however, requires a pest management institution if it is to be implemented.

Given the expected significance of interrelationships between farmers, the frequency of collective decision making in the United States seems rather low. There are a few examples of pest management districts and cooperatives and somewhat more frequent examples of collective decisions reached and enforced with respect to planting dates and cultural practices. Collective action is more likely to take place if there is a clear understanding of the interrelationships, if all parties can benefit without an elaborate compensation mechanism, if and a suitable institution for decision making and enforcement already exists (Olson, 1965). The costs of (1) establishing an organization, (2) determining the interrelationships and optimum management rules, (3) agreeing on how much the gainers should compensate the losers, if necessary, and (4) enforcing the decisions must be less than the benefits from cooperative action for there to be net gains. To date, there are no examples of collective pest management action involving compensation or even situations where it is unclear that all parties benefit every year, and organizations for collective pest control have almost always evolved within or from previously existing institutions.

The benefits from collective action are likely to be more obvious and easier to capture as the interrelationships in the agroecosystem are better determined over time. It also may be possible to reduce the cost of collective organization, decision making, and enforcement with state or federal participation. Further research is needed to determine the potential impact of collective action on pest management and pest use and to determine the best way to establish pest management organizations. If pest management cooperatives or districts can result in reducing the load of pesticides in the environment through more efficient use, then a public subsidy program to offset organization costs may be justified.

Collective pest management strategies may also be more advantageous in the future due to economies of scale in information gathering, processing, and decision making. The development of easy-to-read, perhaps even automatic, insect monitoring devices and computer pest management models could revolutionize pest management decision making. It may be possible to determine optimal pest management strategies for a region nearly as inexpensively as for a single farm. Collective organization may be necessary to capture these gains.

Conclusions

The relationships between the profit-maximizing objectives of the farmer, pest damage, and control costs were explored with respect to their implications to pest management strategies. From this section of the paper, it should be clear that the optimization rules of the economists, especially those which are easy to describe to an interdisciplinary audience, are very difficult to apply to a biological problem as complex as pest management. In turn, the simple management concepts developed by entomologists have become increasingly difficult to "upgrade" with the increased complexity of integrated control strategies and improved understanding of optimization.

The economist can make further contributions by analyzing farmers' attitudes toward risk and their incentives to adopt regional pest management strategies under different institutions such as pest management cooperatives or districts. With respect to risk, the economist can include the trade-offs between risk and higher profits in the benefit–cost analysis of pest management strategies only if the entomologists evaluate the probabilities of different levels of success or failure of the strategy and the opportunity to change strategies in the middle of the program. The interrelationships between the insects and management practices of different farmers in themselves suggest that, in the long run, regional pest management strategies may be desirable if suitable institutions can be designed to gather information, make decisions, coordinate and enforce the actions of the group, and perhaps compensate those individuals who directly lose more than they gain from the regional strategy. Economics will be a valuable science in the design, evaluation, and implementation of social institutions which may be desirable to complement the entomological aspects of pest management.

Literature Cited

Carlson, G. A., 1970, The microeconomics of crop losses, *in: Symposium on Economic Research on Pesticides for Policy Decision Making,* U.S. Department of Agriculture, Economic Research Service, Washington, D.C. (processed), pp. 89–101.

Davidson, A., and Norgaard, Richard B., 1973, Economic aspects of pest control, *Europ. Plant Prot. Org. Bull.* **3:**63–75.

Friedman, Milton, and Savage, L. J., 1948, The utility analysis of choices involving risk, *J. Polit. Econ.* **56:**279–304.

Hall, Darwin C., and Norgaard, Richard B., 1973, On the timing and application of pesticides, *Am. J. Agr. Econ.* **55:**198–201.

Headley, J. C., 1972, Defining the economic threshold, *in: Pest Control Strategies for the Future,* National Academy of Sciences, Washington, D.C., pp. 100–108.

Hillebrandt, Patricia M., 1960, The economic theory of the use of pesticides, Part II, Uncertainty, *Am. J. Agr. Econ.* **24:**52–61.

Hueth, A., and Regev, U. 1974, Optimal agricultural pest management with increasing pest resistance, *Am. J. Agr. Econ.* **56:**543–552.

Olson, Mancur, Jr., 1965, *The Logic of Collective Action,* Harvard University Press, Cambridge, Mass. 176 pp.

Smith, Ray F., 1971, Economic aspects of pest control, Proceedings of the Tall Timbers Conference on Ecological Animal Control by Habitat Management, No. 3, pp. 53–83.

Stern, Vernon M., Smith, Ray F., van den Bosch, Robert, and Hagen, Kenneth S., 1959, The integrated control concept, *Hilgardia* **29:**81–101.

van den Bosch, Robert, 1972, The cost of poisons, *Environment* **14:**18–31.

Willey, Wayne, 1974, Productivity and adoption of pest control technology in California agriculture, Ph.D. dissertation in Economics, University of California, Berkeley. 151 pp.

III

Implementing Pest Management Programs: An International Perspective

William R. Furtick

The developmental stage of a particular agricultural area determines, to a large extent, the type and economic significance of pest problems and the type and extent of pest control efforts. In the more primitive types of agriculture, planned pest control activities are minimal, and the losses to pests are compensated for by planting additional areas to supply the family needs and possibly gain an excess for sale. In essence, enough is planted to satisfy both the pests and the family. In these types of agricultural areas, the long-adapted local varieties usually have an acceptable level of resistance to common diseases and other pests. Weeds are controlled, but seldom adequately, through hand pulling or cultivation.

The so-called "green revolution" is stimulating and accelerating the trend toward modernization of agriculture in many of the developing countries. But it is still significant that outside the more heavily industrialized countries the majority of farmers are at or near the subsistence level, and they do not plan and carry out pest control programs on a systematic basis.

In the more highly developed agricultural areas of the industrialized countries of Europe and North America, there have been a number of distinct patterns in the constantly developing and changing trends in pest control. Prior to World War II, the primary effort was devoted to cultural control measures. This led to widespread use of crop rotations designed to prevent excessive buildup of weeds, diseases, and insects. Scientists sought to understand the life

---

WILLIAM R. FURTICK · Plant Protection Service, The Food and Agriculture Organization of the United Nations, Rome, Italy

cycle and ecology of pests through research, and control programs were designed on the basis of this information to minimize the impact of insects, diseases, and weeds. A number of highly effective programs evolved such as determining Hessian-fly-free planting dates for cereal crops and summer fallow or other tillage programs for control of serious weed pests. Major effort was devoted to the control of plant diseases through breeding for disease resistance as illustrated by the control of various cereal rusts through breeding programs.

With the discovery of highly efficient organic-chemical-based pesticides such as DDT, 2,4-D, and numerous compounds subsequently developed of the chlorinated hydrocarbon, phosphate, and carbamate types, a new phase of pest control was initiated. These new materials were introduced with the close of World War II. Much of the ecological research on pest organisms was diverted to a study of ways to utilize most effectively these powerful new pest control tools.

As a result of high return for a low cost investment, large tonnages of these new pesticides rapidly became the primary tools of the revolution in agricultural pest control on the farms of the developed countries. These coincided with nearly complete mechanization, heavy use of synthetic fertilizers, and use of hybrid and other high-yielding crop varieties. The results led to the familiar recent history of crop surpluses, low commodity prices, and thus cheap food for consumers produced by a highly efficient agriculture.

Along with this incredible achievement of the agricultural sector came signs of pest control problems that have grown more serious and that have prompted reappraisal of the whole approach to pest control. The most serious problems appeared in the control of insects through nearly complete reliance on insecticides. Some major pest species developed resistance to pesticides, and new pest problems developed as a result of the insecticidal kill of parasites and predators that previously maintained the populations of these potential insect pests below the economic injury level.

This general upset of the ecological balance, and the unanticipated side-effects of pesticides on the environment, have prompted the reappraisal and produced a shift in the tactics of pest control. New programs of pest management that integrate a variety of control methods, both new and old, include pesticides as only one element of the control strategy.

Situation and Outlook

Changes on the international scene that started in 1973 and 1974 will have major short- and long-term effects on pest management. Full analysis will require time but the impact from the changes in price and availability of petroleum and other products required in pesticide production will be major. Changes in supply, cost, and ability of many countries to pay for basic manufactured goods, particularly those based upon petroleum, will affect pest management programs. Many of the primary pesticides require petrochemical intermediates for their manufacture, and even those that do not are formulated with solvents and emulsifiers and often packaged in plastic containers derived from petrochemicals. Price increases for pesticides appear inevitable because of increasing manufacturing costs, increasing cost of research on new products, and increasing cost of capital for new investments.

The sharp drop in world food reserves during 1973–1974 resulted in rapid increases in agricultural commodity prices. Higher commodity prices increase the pressure for maximum production and should greatly heighten the interest in the whole field of pest control, in both developed and developing countries. This situation creates the inherent danger of widespread repetition of past mistakes, especially heavy use of pesticides on foreign currency-earning-crops such as cotton in developing countries. If pesticides are available, the same trend would hold for food crops in the developing countries where increased commodity prices may bring pesticides within the purchasing power of many small farmers who previously could not afford them. There is a serious danger that rapid increases in pest control programs in developing areas may outrun the ability of governments to provide sound advisory services. For example, the probable future expansion in the use of pest control measures in developing countries will occur among farmers with the least educational backgrounds. This will impose greater burdens on advisory services than anything previously encountered. Obviously the larger, adequately financed farms owned or managed by the better educated farmers are the first to adopt modern technology. The other should follow at an increasing rate. If there is not a major effort to introduce sound pest management practices along with efforts to intensify production, there will be the potential for disaster. Against this background, it is useful to examine the status of effective pest management schemes within the context of recent and long-standing problems at the national and international levels.

Needs for Development of Effective Pest Management

There has been an increasing amount of discussion on the needs to implement improved pest management systems. The rationale and details of the concept are presented in other chapters, so the focus here will be on the problems of implementation and the steps being taken at the international level.

During the past few years there has been major effort to define and reach agreement on the principles that should govern the elaboration of pest management programs. The Food and Agriculture Organization of the United Nations (FAO), through its "Panel on Integrated Pest Control," has played a leading role in this process. But it is clearly much easier to define the goals and requirements of pest management programs than it is to implement them.

Although there is widespread agreement that the single-method, fixed-schedule approach to the use of pesticides must be replaced by programs that integrate all suitable methods to maintain the pest population below the economic threshold, this change is not easily effected. So far only limited success in making these changes has been achieved in the most highly developed agricultural areas. This stems primarily from the level of knowledge and skill required to develop and implement this approach. Pest management programs have to be developed for each crop and adapted for local agroecosystems. This requires substantial research at the local level. Since there are many factors involved in an agroecosystem, the research is usually rather complex and multidisciplinary. Accordingly, effecting proper control decisions under the variety of agroecosystems in each of the developing countries would be a great challenge. It would require a large number of technicians with adequate training on monitoring, evaluating, establishing economic thresholds and other basic aspects of pest management that are not generally given as part of current plant protection training. In addition, an adequate knowledge base will be required on the control components, including pesticidal, cultural, biological, and the newer nonpesticidal control tactics, including attractants, trapping, and use of pheromones.

The problems that result from overdependence on pesticides are well known, and it is generally agreed that immediate effort should be made to implement integrated pest management systems in areas where these problems are already acute. As agriculture is modernized in developing countries with the accompanying need for more effective pest control, the use of sound pest management programs should be introduced from the beginning in order to avoid the cycle that has characterized pest control programs, such as constantly increasing pesticide use levels, elimination of natural parasites and predators, and resurgence of target species.

If improved pest management programs are to be widely implemented in areas of highly developed agriculture and introduced in the developing world where agriculture is being rapidly modernized, there must be vastly increased funding for research, training and staffing to meet the needs. There must also be major changes in the historic educational and research patterns in the plant protection disciplines. Efficient research and training programs based on an integrated approach cannot be easily accommodated under the rigid disciplinary approaches long institutionalized in centers of higher education and research. Under these historic patterns of institutional organization, entomologists, plant pathologists, weed scientists, and economists conduct disciplinary research and teaching in relative isolation. But when the results of their researches are combined into a management system, there are important interactions that cannot be ignored. The weed-free fields and fence rows that are the pride of the weed scientist may change completely the normal pest–predator–parasite relationships carefully worked out by the entomologist. The microclimate and crop physiology changes introduced by the agronomist through high plant populations and high fertility may completely change the environment for the diseases studied by the pathologist and may lead to unsuspected increases in the reproductive potential of several insect pests. And at the same time, the plant breeder is carefully selecting for the highest yield obtainable under optimum conditions in the nursery where he has made extra efforts to ensure against insects, diseases, and weeds that could reduce the yield.

It may be easier to convert these disciplinary patterns to multidisciplinary teaching and research teams in the developing countries, where institutions are in the developmental stage, than in the established centers of the industrial nations. This will come about in either situation only if major initiatives are taken in this direction by the leaders in plant protection.

The Situation at the National Level

Spectacular cost-benefit ratios have contributed to the tendency for overdependence on insecticides in national pest control programs, but equally important contributing factors have been the relative ease with which farmers can learn to use pesticides and the fact that continued use requires minimal reliance on follow-up guidance. Instead, where carefully planned pest management programs based on economic thresholds are carried out, much more assistance is required by the farmer. Even in areas with highly developed agriculture, the capacity for providing the required services for comprehensive pest management programs has been deficient. It has certainly been beyond the reach of most developing countries to provide such services. In fact the adoption of new technology is occurring in developing countries at a pace which exceeds indigenous institutional capability for staffing the required adaptive research and public service programs. This is where the current crisis exists. The demand from hard-pressed governments for rapid agricultural production increases will be met by inadequate capability at the technical and advisory levels.

However, there have been many positive changes during the past few years at the international level that provide the base for more rapid development of improved national pest management programs. A substantial number of staff from developing countries has been trained abroad, and this number is being increased. Most countries now have qualified local staff although the numbers are inadequate. There has also been rapid improvement in universities and other educational capabilities at the local level. This has been a very important development since expatriates provided the sole source of trained staff in many countries during the colonial era. With their departure, there was a period during which little organizational structure or qualified staff existed. This void was temporarily filled through bilateral and multilateral assistance programs. In many cases, former expatriates were used as the primary source of staff for these international assistance programs. The nature of assistance is rapidly changing as more of the key positions are being assumed by local staff trained abroad and through internationally assisted in-service training programs. There is a trend utilizing highly qualified consultants in support of local staff to maximize use of their overseas training. The normal progression of age is forcing rapid retirement of the original expatriates of the colonial era who have filled an important role in the international assistance programs. As a result of these factors, the whole approach to international assistance is undergoing rapid change. There is urgent need for the development of new approaches.

Fortunately, these needs are coinciding with a period when the rapid postwar expansion of universities and research institutions in the industrialized world has been accomplished, and there are now adequate numbers of trained staff. In many cases, tight budget conditions are causing some reductions in staff which may release trained personnel for international assistance programs.

We are currently in a period characterized by a new spirit of cooperation in plant protection. The establishment and growth of professional societies at the national level, followed by the growth of regional and international organizations, has done much to promote this spirit of cooperation. This has laid a solid base on which to develop the new initiatives that will be required if the world is to feed its growing population.

The Situation at the International Level

The last few years have seen considerable progress in developing the international institutions required for agricultural development. These have been most dramatically brought to world attention through the publicity given to the high-yielding varieties of wheat and rice developed by the international agricultural research centers. These centers have been supported by an international donor group that includes many of the industrial countries, the World Bank Group, United Nations Development Program, the Food and Agriculture Organization of the United Nations (FAO), and foundations such as Rockefeller and Ford. This donor group has formed a loose association called the Consultative Group on International Agricultural Research (CGIAR). They have a Technical

Advisory Committee (TAC) that seeks to identify the most pressing international research needs and proposes institutional structures for solving high-priority research problems. The donors have then shared the cost of the institutions established.

Although excellent progress has been made toward establishment of a global research network, the process is still at an early development stage. Within this network, only minimal attention has been given so far to pest management. On the FAO side, considerable attention has been given to pest management problems, including training, research, and operational aspects.

Much of the regular program of FAO is carried out through the work of constitutional bodies of FAO expert groups, such as the Panel on Integrated Pest Control, the Working Parties on Pest Resistance to Pesticides, Official Control of Pesticides, and Pesticide Residues. The members of these bodies are appointed in their personal capacities by the Director-General of FAO from leaders of the scientific community in their specialities, and they serve for specified terms. These bodies develop recommendations for the Director-General and for member governments of FAO, promote joint research programs, sponsor surveys, and collect and distribute information. Special conferences and other meetings on specific subjects are utilized to bring about an international exchange of views and develop recommendations and issue reports on topics of particular importance to pest and disease management and related programs. As a result of actions taken on recommendations made by the World Food Conference held in Rome in 1974, these activities are being expanded and strengthened.

FAO, as an intergovernmental organization, has been working with and through the scientific community with experts and consultants recruited for this purpose or voluntarily serving in international collaborative programs. An example of one such program is that launched by FAO to obtain reliable information on crop losses due to pests, which is basic to developing rational and economically sound management programs. Four essential steps are recognized in their task: experimental methodology, surveys, integration of surveys and methodology, and the integration of surveys and methodology into insect and disease management systems based on cost–benefit parameters. The FAO Manual on Crop Loss Assessment Methods, periodically updated, provided guidance in these areas.

Some Future Challenges

Much of the world's crop production occurs without the use of any pesticides. Most of the pesticides used in developing countries have been used on plantation crops, but the modernization of agriculture will make it possible for greater numbers of small farmers to use pesticides on food crops. Even though this trend is in process, we still have only limited information on the optimum use pattern for pesticides in pest management programs. In many cases the only source of technical information available to these new users of pesticides will be the pesticide sellers. This calls for closer collaboration between the government and industrial sector. Governments, of course, should assume or continue to bear their share of responsibility for disseminating information on proper usage to both new and experienced users of pesticides. In addition, governments should continue to regulate the types, quality, and use patterns of pesticides at the national level. In the case of many developing countries, outside technical assistance has proven very useful in this respect.

Pest control programs in the past have been considered one of the routine production inputs of a developed agricultural system, and the decisions have been made primarily at the individual farmer level. Because most pests do not respect farm boundaries, it is increasingly clear in many cases that modern pest management systems will require area-wide planning and implementation. This means the decision options will of necessity pass more and more away from the individual farmer. This has been traditional in the public health sector where action programs have been the responsibility of public authorities rather than the individual. This suggests that the same principles might be adapted to pest management for agricultural crops.

Since World War II, an increasing amount of pest control research effort has shifted to the private sector because of the emphasis on pesticides as the primary control agents. With the emphasis now shifting to pest management systems based on use of a variety of control methods that harmonize with changes occurring in the local ecosystem, the research effort of necessity will need to be vastly increased in the public sector.

Expanded basic research on physiology, genetics, and ecology will be required to develop fully new approaches to pest control such as use of pheromones, microbial pathogens, genetic disruptions, etc. These researches will require public funding.

The expanded advisory, monitoring, forecasting, and program planning services required to achieve effective integrated pest management at the local level will also require greatly increased public expenditure. A great challenge in developing and implementing integrated pest management programs will be the integration of not only the disciplines that traditionally have comprised plant protection, but also plant genetics, agronomy, horticulture, and economics. All of these disciplines must collaborate in developing efficient pest management systems.

The complexity of the problem is illustrated by the rapidity with which chemical weed control has become a major factor in pest management. During the past two decades in the developed agricultural countries, the control of weeds has changed from an entirely hand and mechanical operation to one based almost completely on chemicals. As a result the tonnage of herbicides used exceeds the combined use of all other classes of pesticides. The weed scientists are now beginning to encounter problems similar to those already faced by the entomologists. Massive herbicide use has resulted in shifts in weed populations to herbicide-resistant species.

The nearly complete weed control now accomplished with massive use of weed killers is changing the community complex of the agroecosystem with major impact in some instances on the development of both desirable insects and insect pests and on plant disease relationships. These rapid changes have occurred without concurrent study among the weed scientist, entomologist, and pathologist in order to make predictive assessments of the impacts of new technology.

The new capability of herbicides for selective vegetation control, coupled with current increased prices of fuels, farm implements, labor, etc., are stimulating a major reexamination of the tillage requirement of crops. The no-till and minimum-tillage practices will require increased substitution of chemicals for tillage. This trend effects major changes in the ecosystem and thus in the insect and disease relationships. It is imperative that these interactions be studied carefully through a team approach in order to avoid new pest problems that may result from technological change.

Another serious problem in farming systems will be the increased crop vulnerability resulting from reduced genetic variability of crop plants. Undoubtedly market demands (especially those of processing industries) will induce increased genetic uniformity, and this will pose serious constraints on plant breeders and plant pathologists. Another aspect of this same problem is that many of the new high-yielding cereal varieties have disease resistance characters based on single gene factors. Incorporating this type of resistance is fast and easy, but there is the danger of its being easily lost or overcome by a single mutation of the pathogen; consequently, widespread use of such varieties could risk heavy losses. The slower and more difficult process of incorporating horizontal resistance (based on multigene factors) needs urgent attention. FAO has been encouraging more efforts to meet this need in plant breeding programs.

The spread of new high-yielding varieties will bring about increased urgency for the collection and conservation of the large number of indigenous crop varieties that may be eliminated by the rapid spread of the new high-yielding introductions. They are important to long-term pest management programs. These indigenous varieties are genetic reservoirs for disease and insect resistance. FAO and the Consultative Group on International Agricultural Research have already developed new initiatives to counteract this problem.

Serious crop protection problems may arise from the introduction into new areas of threatening plant diseases as a result of increasingly rapid transportation systems and mass movements of people and commodities. The recent introduction of the rust *Hemileia vastatrix* into Brazilian and Parguayan coffee plantations can easily repeat itself in African or Asian rubber plantations with the introduction of the South American leaf blight disease, *Microcyclus ulei,* or with the spread of Moko disease, caused by *Pseudomonas solanacearum,* to the banana or plantain crops of these same regions. The need for better plant quarantines and for investigations on control of these diseases in these regions before a crisis develops is of the greatest importance. New activity is also needed to protect food storage reserves from rodents, birds, and insects.

These are the types of basic programs that need attention and must be coordinated under the new demands for maximum environmental protection coupled with minimized crop losses from pests. These are also allied to the desires of most developing country governments for maximizing rural employment in the process. These requirements may often be in conflict with the type of improved agricultural technology being introduced.

Greatly expanded effort is urgently required to develop the type, quality, and capacity for training the needed staff at all the levels required. This must be done through adequate regional and national training programs. Unless this can be given priority along with assured employment for the trainees at the local level, major personnel shortages will hamper the development of sound pest management in support of an expanding agricultural era.

Continuing attention must also be given to assure the rapid flow of research ideas and results and achieve their full testing, adaptation, and adoption where appropriate. As the diversity of both research and the geographic distribution of research projects increases, the problems of ensuring rapid information flow become increasingly difficult.

Encouragement, possibly financial support, must be given to the industrial sector to assure continuation and expansion of the research required to discover, produce and market pesticides that will be more specific and safe. The industry faces this challenge in a time of spiraling costs caused by the increased research to ensure environmental protection and increased cost of petroleum products. Serious study is required on how these costs are distributed between the industrial and the public sector. Perhaps an effort should be made to develop an international convention on registration requirements to prevent unnecessary cost.

Little attention has been given to socioeconomic research to ensure that the technology developed is appropriate and acceptable to the local farming community. In particular, financial and institutional arrangements must be developed to ensure that pest control materials are available as required to peasant farmers in developing countries.

Maximum cooperation and coordination is imperative if the global requirements of pest management, as a critical component of food production technology, are to be met in the future.

IV

Pest Management: Principles and Philosophy

Edward H. Glass

Animal and plant populations in natural biotic communities are relatively stable and outbreaks of any one species are rare (MacFadyen, 1957). Furthermore, most species exist at relatively low population levels. In these natural communities, most animals are classified as herbivores, parasites, or predators; few species are truly saprophytic (Pimentel, 1969). Ecologists refer to this phenomenon as species equilibria in community structures.

Such equilibria have evolved over long periods of time. Populations are controlled by many biotic and abiotic factors, including for example, both inter- and intraspecific competition, environmental heterogeneity, natural enemies, and weather. I do not propose to elaborate on the complex nature of these several mechanisms for they have been extensively reported in the literature and are not essential to this discussion. Even though species population oscillations occur, their magnitude seldom reaches destructive levels. Thus, in natural communities, pests, as we know them in agricultural crops, seldom exist. When some disruption occurs, such as a climatic change or an introduction of a new species, the steady equilibrium is disturbed and numerous population fluctuations may be anticipated until a steady state again evolves.

The development of agriculture over the last 10,000 years or more has had great impact on man and his environment. Among other consequences, it has directly or indirectly produced an abundance of food and fiber, has led to a human population explosion, has led to the destruction of vast areas of forests, prairies, and other wild habitats over most of the world's arable land, and obviously has disrupted on a worldwide scale many of the species equilibriums

---

EDWARD H. GLASS · Department of Entomology, New York State Agricultural Experiment Station, Cornell University, Geneva, New York 14456.

described above. In developing agriculture we have created crop pests as we know them today. Let us explore briefly the progress of agriculture and how it has impinged on crop protection problems.

During the gradual evolution of agriculture, there have been a number of known severe pest outbreaks and unquestionably more occurred but have been lost in the mists of unrecorded history. There have been recurrent locust plagues dating back at least to biblical times, there was the potato-blight-induced famine in Ireland in the 1840s, and the greatest crop loss in history—$1 billion, more or less, of corn in the United States—was caused by the corn leaf blight in 1970. In spite of these and many other outbreaks, the agricultural enterprise has been very effective in producing food and fiber for a rapidly expanding population. There are, however, some good reasons to be concerned and even alarmed at prospects for successful future crop protection.

During the gradual slow evolution of early agriculture, man evolved relatively stable agroecosystems. He did this more by necessity and chance than by design. He obviously harvested those plants which survived pest damage and thereby unconsciously selected for pest resistance. He practiced agriculture in discrete communities with little exchange of crops or their pests from one area to another so that there was much genetic diversity throughout the geographical range of any particular crop. He operated on a small scale, and environmental disruption was minimized. Thus in primitive agriculture, a measure of species stability was maintained.

But even this degree of stability was disrupted as a result of the explorations and development of the world's land masses that was begun so intensively in the fifteenth century and proceeded with ever-increasing intensity into the twentieth century. As the Americas, Africa, Australia, and the large islands of the world were explored, exploited, settled, and developed, plants and animals were transported across previously impenetrable barriers. Maize and potatoes were brought from the Americas to the rest of the world; cereal grains, edible legumes, and the deciduous tree fruits were carried from Europe and Asia to other continents. Hundreds of plant and animal species have thus been spread throughout the world.

When crop plant species were carried into new regions, many of their pests were also brought along with them, usually without their normal complement of biological control agents. These often became more severe pests of their original host in the new habitat than in the old and frequently found indigenous species to be suitable hosts. Well-known examples are the codling moth, the cottony cushion scale on citrus, the Hessian fly, Dutch elm disease, and the chestnut blight. Further crop protection problems developed as native species, often considered to be innocuous, adapted to introduced crops. The apple maggot, a little-known insect infesting *Crataegus,* adapted to apple and is now considered the worst enemy of this fruit in northeastern United States and southeastern Canada (Dean and Chapman, 1973).

The agricultural revolution of the twentieth century has further disrupted and complicated the stability of pest species in our agroecosystems. Through application of scientific genetic principles and technology, new high-yielding cultivars adaptable to mechanized culture, harvesting, and postharvest procedures have been selected and intensively planted over wide areas. Many new cultivars are genetically uniform and potentially more vulnerable to pest epidemics. The corn leaf blight epidemic of 1970 represents a case in point (Horsfall, 1972). The new cultivars must frequently be protected from pests, and so pesticides are extensively required to attain potential yields. Many cultural practices of modern agriculture also enhance susceptibility to disease, weeds, nematodes, or insects. These include: (1) irrigation, which favors many disease and insect pests as contrasted to the fluctuating soil moisture levels under natural rainfall conditions, (2) multiple cropping, which promotes rapid pest population increases, (3) dense crop plant populations, resulting in environmental changes favoring some pests, and (4) fertilization, which produces larger and more succulent plants, which are often more susceptible to pest attack than those grown in low-fertility levels (Smith, 1972).

Rice production in the Philippine Islands offers an excellent example of what happens when production is intensified quickly without adequate safeguards against pest epidemics. Traditional rice culture in that country produced each year a modest but rather constant yield. The varieties were rank tall types that survived on low fertility and competed successfully against weeds. They were not immune to insects, diseases and rodents, but all pests were reasonably tolerated. Rice was cultivated once a year during the wet monsoons and was followed by a 5- or 6-month fallow through the dry season. Pest survival during the dry period, when hardly a green blade of rice or other grass can be found except along stream banks, was low, and only insignificant populations survived to attack the next crop.

Compare this situation to the new intensive rice culture in the same country. First, there is irrigation so one crop follows another throughout the entire year. The new varieties are short and stiff and require weed control to realize full yield potentials. The earliest high-yielding varieties were not selected for pest resistance. Inevitably pest problems increased significantly and in 1971 severe losses from leafhopper attack and the leafhopper-transmitted virus disease "tungro" were experienced over thousands of hectares in the "rice bowl" of Luzon. (Glass *et al.,* 1972). More recently another leafhopper-transmitted virus, grassy stunt, has struck the area. I observed this change in rice production technology ("green revolution") as it began in 1967 and again in 1971 when the pest epidemics were in progress.

My intent in citing the above example is not to be critical or suggest the discontinuance of intensive agriculture in the Philippines, the United States, or elsewhere. We must sustain and expand intensive production to feed the expanding world population (Borlaug, 1971). The rice example does illustrate, however, the seriousness of the pest potential in intensive modern agriculture, the complexities of interactions between practices, and the potential counterproductiveness of the unilateral implementation of disciplinary inputs of modern agriculture. It dramatizes the need for an interdisciplinary approach to crop production.

Traditional Crop Protection Procedures

Detailed accounts of the history and development of crop protection methods are thoroughly documented in numerous publications and need not be discussed here. Further, we need not be concerned with the primitive wooden flails used in futile attempts to protect grain crops from locust plagues or slingshots used to drive away flocks of grain-devouring birds or even the handpicking of insects. Instead, we shall direct our attention to the twentieth-century efforts that have evolved along with the modern agricultural revolution.

Our major twentieth-century efforts in crop protection have been simplistic, i.e., the unilateral use of one or another crop protection tactic. We have sought to find quick, effective procedures to solve pest problems. We have used cultural methods, planting of dominant monogenic resistant crop types, the release of biological control agents, and the application of highly effective chemical pesticides. We have used fly-free planting dates to prevent damage to wheat by the Hessian fly, rust-resistant wheats to avoid losses by this pathogen; we have introduced the Vedalia beetle to control the cottony cushion scale on citrus; we have employed lead arsenate, DDT, organophosphate and carbamate insecticides to protect pome fruits from the ravages of the codling moth, crop rotations to reduce soil nematode populations, and many more methods. Each of these has been effective, at least temporarily, and some have had spectacular success. But these have brought problems. The fly-free dates do not coincide with agronomically optimum planting periods; new pest biotypes evolve and attack resistant cultivars; citrus pests other than cottony cushion scale develop or are introduced and these pests require the use of chemicals which destroy the introduced scale predator; the codling moth evolves strains resistant to one chemical after another; and many crop rotations simply do not fit into the highly competitive modern agricultural systems.

Other types of problems are involved with these crop protection procedures. For example, there are environmental, societal, and economic factors to be considered. These loom larger and larger as human populations explode, the demand for food increases, and the urban–rural requirements compete for land, air, and water. The period of unrestricted activities and expansion of segments of our society is fast drawing to a close. The increasing restrictions on the use of pesticides is dramatic evidence of such impact on crop protection activities.

The overview of crop protection problems presented by Furtick in chapter III is ample evidence of the need for an improved approach to the problem. Those of us who have experienced many of the successes and frustrations of the past four decades are firmly convinced of this need. We must develop systems of crop protection that will establish satisfactory species equilibria in our modern intensive agriculture. These must be at economically sound population levels and must be consonant with environmental and societal requirements.

An interdisciplinary pest management approach to crop protection offers much promise for achieving such objectives. Let us now examine the principles and philosophy of this approach.

Pest Management

The meanings of several terms used in crop protection have changed during the past decades; the result is some confusion. A few definitions of these terms as used in this chapter should eliminate possible confusion. The term *pest* is simply defined as any organism detrimental to man, whether it be an insect (formerly the term was restricted to insects and related organisms), disease organism, weed, rodent, or other. It is obviously an anthropocentric concept. Some who view the survival of our total biotic system above immediate human interests consider man to be a pest of the world (Corbet, 1970). *Pesticides* are substances used to control pests; thus insecticides, fungicides, nematocides, avicides, etc., are all pesticides. *Pest control* is a broad term that can be applied to any procedure employed to reduce pest populations or prevent their detrimental effects. *Pest management* designates a philosophy and methodology of restricting pest numbers to noninjurious levels (Huffaker, 1970). *Integrated control* is now most commonly used in the United States interchangeably with *pest management* and refers to an integration of control tactics into a strategy of pest control. Formerly it was applied by some only to the integration of biological and chemical controls. I shall devote the remainder of my time to the philosophy and principles of pest management as they apply to crop protection.

The FAO panel of experts on integrated pest control defined ''integrated control'' as ''a pest management system that in the context of the associated environment and the population dynamics of the pest species, utilizes all suitable techniques and methods in as compatible a manner as possible and maintains the pest populations at levels below those causing economic injury.'' (FAO, 1967). This definition incorporates the concept of pest management as defined by the Entomological Society of America and most authorities. It is in this sense that I use the terms ''pest management'' or ''integrated pest management.'' Thus, *pest management* means a systems approach that encompasses not only the immediate objective of preventing pest losses but also consideration of long-term objectives with regard to economics, society, and the environment.

The integration of control tactics has been practiced long before the philosophy and guiding principles of pest management were formulated. Entomologists Michelbacher in California and Pickett in Nova Scotia were early practitioners of the integration of biological and chemical control in the late 1940s. The real impetus for pest management, however, was provided by the problems that developed in the early 1950s with the unilateral use of pesticides. These were development of resistance, resurgence of pests following chemical treatments, destruction of beneficial insects, creation of new pest problems, effects on nontarget organisms, environmental contamination, and others. Many disciplines have been involved in developing the pest management concept, but entomologists have been and continue to be leading exponents both nationally and globally. While entomologists may have been early proponents of pest management, scientists in other crop protection disciplines have also been practicing integrated control and have shared in the development of the rationale and implementation of the concept.

The concept of pest management implies a manipulation of the agroecosystem in such a way that pests are maintained at subeconomic population levels. The goal is to incorporate into our crop production systems those components and practices which are needed to dampen the oscillations of pest populations so that the upper levels are noneconomic. We must do in a short time what nature can evolve only after extended periods of time. The ecological aspects of these manipulations will be elaborated in subsequent chapters. Here, I shall attempt to provide a general background of the methodologies, techniques, and strategies useful in pest management. Space limits a comprehensive, detailed review, but such may be found in the literature (Huffaker, 1972; USDA Symposium, 1966).

Tactics in Pest Management

Most of the available tactics or techniques useful for integrated control of crop pests are not new even though some may seem so to those who were educated in the decade following World War II, when heavy reliance had been placed, particularly by entomologists, on chemicals to the near exclusion of other methods. Some tactics are simply the optimization of naturally occurring phenomena (host resistance and biological control), while others are artificial (cultural and chemical). Each is more or less feasible and sound on pragmatic, economic environmental and social grounds. Thus a herbicide may introduce a chemical into the environment, tillage may result in excessive soil erosion, and hand weeding may be uneconomical. Some tactics useful in pest management are discussed herewith.

Plant Resistance. The development and use of crop cultivars resistant or tolerant to one or more of its pests is economically and environmentally sound. It has been a major element in the control of certain nematodes, plant pathogens, and a few insects. For example, there are tobacco cultivars resistant to six major fungal, bacterial, viral, and nematode pests. Hessian-fly-resistant wheat is perhaps the most important instance of insect resistance. Entomologists in particular are just beginning to recognize the important role of even relatively low levels of plant resistance in a total program to manage pests. Even though few attempts have been made to improve through breeding the ability of crop plants to compete with weeds, there is evidence to suggest that this is a potentially profitable approach. Development of resistant cultivars is a long, expensive procedure but it provides agriculturists with an economical and environmentally sound tool for pest control.

Cultural Controls. This is one of the oldest methods of crop protection. Early or late planting, cultivation, fallowing, sanitation, rotations, and other methods have been important in crop protection. Many of these have been abandoned in recent years because other methods more compatible with crop production practices and needs became available. Now cultural controls are being reexamined for their usefulness in pest management since they are compatible with most other control tactics.

Any cultural control operation may influence other pests, either favorably or adversely. Elimination of weeds, for example, may remove a source of a crop pathogen or insect pest or may eliminate a source of parasites and predators of pests. The alteration of cultural practices for any reason may have important influences on pest populations and damage. Therefore, interdisciplinary considerations through a team approach are absolutely essential.

Biological Control. The important role of biological control organisms in regulating pests, especially insects, of agricultural crops was not fully realized until they were destroyed by applications of chemicals harmful to the beneficial forms but not their hosts. For example, DDT applied to apples controlled many important pest species, including codling moth and apple maggot, but was not effective against phytophagous mites. The parasites and predators which had previously regulated mites were destroyed, and populations exploded. Many examples could be cited; some will be discussed in detail in later chapters.

The tremendous importance of biological control agents for regulating crop insect pests and perhaps to a lesser extent for other types must be recognized, and all possible effort must be made to enhance their effectiveness. Considerable progress has been made in recent years to integrate chemical and biological control but much remains to be done.

Pesticides. These control agents have been, are now, and will continue to be, in the foreseeable future, basic tools in pest management. In fact, there are pest situations for which there are no known alternate management methods. They provide a dependable, rapid, effective and economical means of controlling whole complexes of crop pests. They have been substituted for other more cumbersome or expensive methods such as crop rotation.

For all the good points they possess, however, problems in their use are expanding and intensifying as has been stated previously. Problems with resistance are most acute for insecticides, but cases of resistance development have been found in plant pathogens, weeds, and rodents. Increased restrictions on use of certain pesticides and uncertainty about the future discovery and development of replacement materials are alarming.

Other Tactics. There are several other old and new techniques available that have application in pest management. Autocidal methods, such as the sterile male technique, are usually thought of as a means of eradication, but they can be used for management of certain insect species. Insect pheromones are potent tools for use in conjunction with other methods and even for direct control by trapping or mating inhibition. Certain plant growth regulators are useful as herbicides and insect growth regulators have promise of being useful for regulating certain insects and perhaps other pest types.

We should also mention quarantine, eradication, and regulation as tools in pest management. The most effective method for control of exotic pests is to prevent their introduction, an increasingly difficult goal. In case introduction occurs, eradication may be the optimum approach if there is early detection and elimination is feasible. There are several instances of success in eradication, but there have been many failures and good judgment is needed. Regulation is another important tool for managing many pests, including weeds, disease organisms, and insects. Use of weed-free seeds and disease- and insect-free plant propagules is an excellent control procedure and can be attained by regulation. Regulation of pesticide use could also be useful in preventing misuse of pesticides in pest management systems.

Strategies in Pest Management

There are three basic strategic approaches to pest control: (1) complete reliance on natural forces, i.e., no overt action, (2) prevention or eradication, and

(3) containment or correction. The first is not a practical strategy for the complex of pests found in most agricultural situations. However, it may be most appropriate for one or more species among the complex. The most productive strategy is determined by a number of considerations. For example, preventive chemical control applications are the only practical approach to controlling apple maggot and apple scab in New York. Failure to prevent infection results in intolerable losses and increased pesticide use just to prevent further losses. In fact, we have a saying that "trying to permit a little maggot or scab is as impossible as to become only a little pregnant." On the other hand, there are situations where pest attacks are irregular, are tolerable at low levels and may be successfully contained or corrected when they do occur. Most aphids, mites, and many plant diseases can be managed successfully by this strategy.

The overall philosophy of pest management employs the strategy of maximizing natural control forces, i.e., natural enemies and plant resistance, by utilizing any other tactics with a minimum of disturbance and only when crop losses justifying action are anticipated. Therein lies a major problem of pest management; how to predict or anticipate economic losses and how to determine economic thresholds for individual pests and particularly for pest complexes. The formulation of economic thresholds is a complicated process and more information is needed on the economic aspects of pest control, especially with regard to benefits and hazards, research alternatives, and social strategies.

The integration of control practices must be based on the realization that individual pest species are single components of a complex agroecosystem and that interactions among the components cut across the artificial lines created by the taxonomically oriented scientific disciplines involved in crop protection. Therefore, the development and implementation of integrated pest management requires both a disciplinary and an interdisciplinary approach. Entomologists, weed scientists, nematologists, and plant pathologists broadly knowledgeable about pests and their control must address themselves to the concept of integrated control. Modern computer technology and "systems analysis" provide a means by which the resulting tremendous volume of complex information from the several disciplines can be integrated and synthesized into a practical yet viable strategy.

Agroecosystem analysis and modeling have two essential values in pest management: (1) systems analysis helps to identify areas where additional information is required, and (2) predictive models for crops and pests will enable specialists to determine more accurately economic thresholds and predict if and when they will be reached. Modeling will play a key role in the development of pest management strategies; however, the development and implementation of integrated control can and is proceeding without the complex models which eventually will be so valuable.

Implementation of Pest Management

Knowledge of an agroecosystem, its pests, the complex interactions that regulate these, and the development of an integrated system of pest management within that system are not enough. Success is attained only with implementation, a procedure which requires a technology of its own. Acceptance of pest management systems by farmers and others has been slow for a number of reasons—insufficient funds, lack of understanding, conflicts of interest and traditions, among others. I anticipate fewer and fewer problems of implementation as management systems are improved and their value to the agricultural producers and society are more widely recognized.

Limitations of Pest Management

While we recognize the great potential of pest management for protecting our crops from the ravages of pests with systems which are consonant with maintenance of a viable, productive agriculture without intolerable disruption of the environment or society, we must also realize that there are limitations. A major current limitation is the lack of information about our pests and the agroecosystems. Much more detailed knowledged is needed for the integration of control tactics into an overall system than for direct control of individual pests. Lack of information often leads to misunderstanding of the goals and tactics of a program. Another limitation is the time and expense involved in developing integrated approaches. And finally there are certain problems for which pest management does not presently seem to have application.

Summary and Conclusion

In summary, agriculture, particularly that involving the intensive high-yielding agroecosystems of the industrialized nations, has disrupted the species equilibriums that occur in natural communities and to a lesser extent in primitive agriculture. The introduction of high-yielding, genetically uniform varieties without regard for pest susceptibility; use of fertilizers, irrigation, high-density plantings, monocultures; extensive transport of crops; and other practices have created agroecosystems which on balance favor pest epidemics and repress natural controls. Adequate attention has not been devoted to the problems of providing protection of our new agriculture systems.

The rapid rate of changes being made in agricultural practices today and the worldwide demand for all the food we can produce do not allow the time necessary for crops, their pests, and biological agents to reach species equilibriums by "natural" means. The approaches to crop protection made during the past few decades have had many great successes and some great failures. These approaches have been largely simplistic, unilateral applications of such control tactics as pesticides, plant resistance, and introduction of biological agents. Most have not been based on sound biological, ecological, environmental principles. As pesticide-resistant strains and new biotypes able to attack formerly resistant crop cultivars have evolved, as environmental problems mount, these unilateral approaches are less and less acceptable and do not appear to be practical approaches for the extended future.

The problem is to establish as quickly as possible species equilibriums in our agroecosystems wherein the upper oscillations of pest populations are maintained below economically damaging levels. Pest management offers the most promising approach to achieving this goal without disrupting agricultural production systems, the environment, or society. It is an interdisciplinary ecological approach employing a philosophy and methodology of restricting pest numbers to noninjurious levels.

Literature Cited

Borlaug, N. E., 1971, Mankind and civilization at another crossroad, 1971 McDougal Memorial Lecture, *in:* Food and Agriculture Organization Conference, Rome, November, 1971.

Corbet, P. S., 1970, Pest management: objectives and prospects on a global scale, *in: Concepts of Pest Management* R. L. Rabb and F. E. Guthrie, eds., North Carolina State University, Raleigh, pp. 191–208.

Dean, R. W., and Chapman, P. J., 1973, Bionomics of the apple maggot in eastern New York, New York State Agricultural Experiment Station, Geneva, *Search* 3(10). 64 pp.

Food and Agricultural Organization (FAO), 1967, Report of the first session of the FAO Panel of Experts on Integrated Pest Control, Rome.

Glass, E. H., Smith Jr., R. J., Thomason, I. J., and Thurston, H. D., 1972, Plant protection problems in Southeast Asia, Report for U. S. Agency for International Development, University of California Contract No. AID/csd-3296.

Horsfall, J. G., 1972, Genetic vulnerability of major crops, National Academy of Sciences. 307 pp.

Huffaker, C. B., 1970, Summary of a pest management conference—A critique, *in: Concepts of Pest Management* (R. L. Rabb and F. E. Guthrie, eds.), North Carolina State University, Raleigh, pp. 227–242.

Huffaker, C. B., 1972, Ecological management of pest systems, *in: Challenging Biological Problems, Directions Toward Their Solution* (John A. Behnke, ed.), Oxford University Press, New York, pp. 313–342.

MacFadyen, A., 1957, *Animal Ecology,* Pitman, London.

Pimentel, D., 1969, Animal populations, plant host resistance and environmental control, *in:* Proceedings of the Tall Timbers Conference on Ecological Animal Control by Habitat Management, No. 1, pp. 19–28.

Smith, R. F., 1972, Management of the environment and insect pest control, *in:* 1972 Food and Agriculture Organization Conference on Ecology in Relation to Plant Pest Control, Rome, December, 1972.

U.S. Department of Agriculture, Symposium, 1966, Pest control by chemical, biological, genetic, and physical means, ARS 33-110, July, 1966.

V

Pest Management in Ecological Perspective

Philip S. Corbet

The Parable of the Beans and the Bean-Eaters
Suppose you try to make a very simple system—a field, for example, in which you have killed everything in order to grow just one kind of living thing, say beans. . . . You will find that all sorts of plants and animals—mainly fungi and insects—try to come (to) eat your beans.

M. STRONG (1973)

Pest Problems: Their Nature and Causes

Pest management is concerned with the suppression of pests and with the alleviation of pest problems. The concept of a ''pest'' has meaning only in a human context: a pest is an organism that *man* regards as harmful to his person, property, or environment. As Rudd (1971) has observed, the word ''pest'' (like the word ''weed'') is defined only according to its impact, direct or indirect, upon man, often in environments that man himself has modified considerably. Man makes an organism a pest as soon as he requires something it needs and which he is not prepared to share with it; and he frequently makes it a ''worse'' pest by manufacturing an environment that favors its increase and survival. Since pests, by definition, are competitors for resources that man wishes to preempt for his own use, the more resources he tries to preempt, the more pests he will encounter. Thus a study of pest management in ecological perspective

---

PHILIP S. CORBET · Department of Biology, University of Waterloo, Waterloo, Ontario, Canada. Present address: Department of Zoology, University of Canterbury, Christchurch, New Zealand.

must take account of the activities of man—the species that creates and aggravates pest problems and that devises treatment designed to alleviate them.

When seeking to identify the causes of current crop-pest problems it is useful to distinguish between the creation and the aggravation of such problems. We may suppose that all crop-pest problems owe their *creation* to the existence of agriculture: the persistent holding back of biotic communities in very early, immature stages of ecological succession—stages in which the population numbers of the component species are least stable. It is of agriculture that Wylie (1968) speaks when he says: "to open up lands in our fashion is to shut down their steady state."

Having been created by agriculture, pest problems are then *aggravated* by the way in which it is now being practiced—a circumstance noted in earlier chapters. In today's intensive, industrialized agriculture, pest problems can be made worse by standard procedures for cultivation and for crop protection. The standard procedures for cultivation are designed to produce individual plants of greater size or yield (by increased use of irrigation, fertilizer and high-yield varieties) and to produce a greater number of plants per unit area (by intensive monoculture) or per unit time (by multiple or continuous cropping). Such practices aim to provide an unusually rich, dense, continuously available substrate that accordingly favors the rapid reproduction of those insects that can use it for food. Crop-protection procedures are of course designed to secure the highest possible proportion of this substrate for man. They typically involve the application of herbicides and pesticides which reduce the species diversity of the agroecosystem; this decreases the mortality caused by the parasites and predators of pests and thus renders more likely the assumption of pest status by organisms that were not regarded as pests before. And should any pests have acquired resistance to the pesticide being applied, such reduction of mortality can lead promptly to serious pest outbreaks.

These current procedures associated with cultivation and crop protection—all of which can aggravate pest problems—are themselves direct consequences of the increasing pressure that is being placed on arable land. It is therefore unlikely that there will be a significant, overall reduction in the number or severity of crop-pest problems until this pressure on arable land can be relieved. Persisting in our search for the main causes of pest problems, we must now ask: "To what is this pressure due?"

In the first place it reflects the fact that arable land is a finite resource, or more correctly, that it is now a *diminishing* resource: most of the world's easy-to-develop fertile soils have already been brought into agricultural use (Brown, 1967; Barlowe, 1972) and much arable land is lost each year to nonagricultural uses (see Watt, 1973).

Secondly, the pressure is caused by the escalating demand for increased yields per acre. This demand is powered by what may be called human "need" and "wants." The "need" is for *more* food, required for subsistence, or at least survival, by increasing numbers of people as the human population grows. The "wants," which reflect people's rising expectations, are for *better* food (i.e., more animal protein) and for profit, since agricultural produce has now become a commercial commodity and not just a source of food (see Mead, 1970); indeed, as land becomes scarcer and agriculture more mechanized and capital-intensive, farmers must make a *rising* profit simply to obtain an adequate return on investment.

The causal relationships identified in the preceding paragraphs favor the conclusion that today's crop-pest problems owe their genesis (in the main) to four separate developments in man's history:

(a) The development of agriculture, involving the conversion of ecosystems from natural to artificial productivity and the provision of a rich, uniform food substrate, for man and pests.
(b) The development of economics, involving the use of surplus food as a medium for trade.
(c) The development of preventive medicine, involving a reduction in human mortality (without a compensating reduction in natality).
(d) The application of a high-energy technology to agriculture and its supporting activities, which involves a short-term increase in productivity based on subsidies from fossil fuels and mineral ores.

It is instructive to view these four developments in ecological perspective. First, we may note that all are extremely recent: more than 99% of man's known existence as a species has been preagricultural. Second, all are without precedent or analogy among other organisms. Accordingly, they must be regarded as ecological anomalies. As Wylie (1968) observes: "We are a young experiment." But although the developments themselves constitute ecological anomalies, their results do not. Indeed, they are only too familiar to pest managers! For together these developments have provided the two conditions needed for a pest outbreak to occur (Clark *et al.*, 1967): (1) a lasting increase in the supply of food or some other previously limiting resource and (2) a lasting decrease in the frequency or severity of repressive interactions (e.g., disease-induced mortality) that previously limited exploitation of the resource base.

Thus can we bring into focus man's outbreak and use of technology as causes of what Southwood (1972) has so aptly named "the environmental complaint"—the deteriorating condition resulting from man's treatment of the biosphere. Pest problems are among the many symptoms of this affliction and so will persist as long as the "complaint" is allowed to continue. From an ecologist's viewpoint there is no immediate prospect of its continuation being checked by purposeful human intervention because current policies followed by the relevant international (United Nations) agencies favor the maintenance of the two conditions that brought it about: the Food and Agriculture Organization is committed to increasing the substrate that is permitting population growth; and the World Health Organization is committed to reducing mortality. It appears then that the four developments, those ecological anomalies referred to earlier, have generated trends that are leading to increasing instability and that do not appear to feature built-in corrective mechanisms, at least not ones that modern man would regard as acceptable. Our four formidable developments have much to answer for! And so, therefore, has that anomalous biological event that made them possible: the disproportionately great ability of one species, man, to rationalize and communicate what it experiences, and then to use such knowledge to modify its environment.

Accordingly it may be said that, to a biologist, the root cause of the environmental complaint (and so also of its several symptoms, including crop-pest problems) is that one of the recent primates has a peculiar kind of brain: a brain that has allowed it to modify ecosystems in such a way, and to such an extent, that the mechanisms that had previously kept its density below the carrying capacity of the environment were rendered ineffectual before this species had devised workable replacements for them. It is this brain that enables man to undertake projects that have great impact on the biosphere (such as agriculture, preventive medicine, and atomic physics) and yet not to foresee, or allow for, their long-term effects. "Modernity has added to man's knowledge of the natural world and his relationships with it without giving him an adequate philosophy to guide these relationships toward dynamic stability and ecological renewal" (Caldwell, 1972). In short, the human brain enables man to *generate* environmental problems but not necessarily to *solve* them. Well may McHarg (1969) ask whether the contents of the human skull should be regarded as the apex of biological evolution or as a threat to human survival! And well may we reflect again (as ecologists) on the adaptive value of being simple-minded (see Smith, 1952; Robin, 1973), and on the pertinence of that earlier sanction against eating of the fruit of the tree of knowledge!

From the wisdom born of hindsight, we can view man's contemporary shortcomings without surprise. His biology was shaped in the hundreds of thousands of years of the preagricultural period during which he lived as a gatherer and hunter (Washburn and Lancaster, 1968). The cultural developments that have occurred during the last ten thousand years have profoundly altered his way of life but have not yet provided enough time for much genetic change: "man is facing the space age with a genetic equipment determined largely by the selection pressures which produced his pre-neolithic ancestors" (Barnes, 1970). I suspect that James Branch Cabell's ingeniously illogical pronouncement (see Hubbert, 1969) is even more appropriate now than when he made it in 1926: "The optimist proclaims that we live in the best of all possible worlds; and the pessimist fears this is true"!

Pest Problems: Their Possible Solutions

It is not an encouraging scenario that has emerged from this analysis: a preneolithic man struggling with a runaway outbreak of his population and relying for subsistence on a stressed substrate whose productivity has been temporarily raised by an injection of nonrenewable resources. Quite so; but if this scenario *is* realistic (as I believe it to be), then the ecological perspective that revealed it may also provide the understanding needed to identify a goal that is ecologically sound and perhaps also to perceive and articulate actions that might achieve that goal.

The ecology of nonhuman organisms living in habitats unmodified by man and the ecology of nonagricultural man suggest that an acceptable goal for *Homo sapiens* (regardless of his culture) would be the attainment of long-term stability at a density somewhat below the sustainable carrying capacity of the biosphere for man. Some might wish to add the rider that, to be worth achieving, this goal should be attained without human populations being subjected to large-amplitude fluctuations or to major cultural discontinuities.

If this goal is to be achieved it is evident what must *not* be done: further increase of human numbers and, in due course, of human food will be counterproductive. Less obvious are the actions that *will* be required if the goal is to be achieved. One may suppose that one of the first would be a thorough review of man's prevailing attitude toward resources, particularly agricultural resources, with a view to relieving the pressure on arable land (see Corbet, 1973). After having grown accustomed during the last few centuries to operating as an "*r*-strategist" (see MacArthur and Wilson, 1967), living on capital, with expansion and dispersal as goals, man will not find it easy to revert to the life-style and attitudes of the *K*-strategist, who lives on income, and for whom attainment of equilibrium is the main objective. Such a change would involve the conversion of *Homo sapiens* "from conqueror of the land-community to plain member and citizen of it" (Leopold, 1949). To accomplish this change there will first have to be accelerated growth of ecological awareness among the public and its elected representatives, accompanied perhaps by some moral and legal means of "mutual coercion, mutually agreed upon by the majority of people" (Hardin, 1968). In short, people must be quickly brought to recognize the nature and causes of the environmental complaint and the consequences of permitting it to continue untreated.

There are some who hold the view that such curative action does not lie within the discipline of pest management or within the responsibility of its practitioners. Others contend that persons who (like pest managers) are employed to treat the *symptoms* of the environmental complaint could be expected to assume some responsibility for treating its *causes* also. Certainly pest managers are better qualified than most scientists to understand and explain the principal components of man's current ecological predicament.

The discipline of pest management has probably never occupied so important a position as it does today: many now acknowledge that there are not nearly enough trained, capable people to tackle current and prospective problems in this field. And yet, despite this manifest imbalance, one feels that if all pest managers devoted even 1% of their professional time to improving public awareness of the *causes* of environmental problems and of the need to start working toward long-term objectives *now* (rather than tomorrow), perhaps some prospect would emerge of solving these problems in a lasting way; of creating an environment in which pest problems would diminish instead of continuing to increase; and of generating a political climate in which crop-protection measures were decided with reference to the ecological, rather than the economic, injury level; and in which patterns of resource consumption reflected a desire to reduce the gross national cost (Boulding, 1971) rather than to increase the gross national product. I believe that in such an environment the resources available to the discipline of pest management could serve mankind even better than they do now.

Those who accept this responsibility may find their resolve strengthened by this stirring directive from McHarg (1968): "We must abandon the self-mutilation that has been our way, reject the title of planetary disease which is so richly deserved, and abandon the value system of our inheritance which has so grossly misled us. We must see nature as process within which man exists, splendidly equipped to become the manager of the biosphere; and give form to that symbiosis which is his greatest role, man the world's steward."

Literature Cited

Barlowe, R., 1972, *Land Resource Economics* (2nd ed.), Prentice-Hall, Englewood Cliffs, N.J.

Barnes, H. F., 1970, The biology of pre-neolithic man, *in: The Impact of Civilization on the Biology of Man* (S. V. Boyden, ed.), University of Toronto Press, pp. 1–18.

Boulding, K. E., 1971, Environment and economics, *in: Environment, Resources, Pollution and Society* (W. W. Murdoch, ed.), Sinauer Associates, Stamford, pp. 359–367.

Brown, L. R., 1967, The world outlook for conventional agriculture, *Science* **158:**604–611.

Cabell, J. B., 1926, *The Silver Stallion*. McBride, New York (quoted by Hubbert, 1969).

Caldwell, L. K., 1972, An ecological approach to international development: Problems of policy and administration, *in: The Careless Technology* (M. T. Farvar and J. P. Milton, eds.), Natural History Press, New York, pp. 927–947.

Clark, L. R., Geier, P. W., Hughes, R. D., and Morris, R. F., 1967, *The Ecology of Insect Populations in Theory and Practice*, Methuen, London.

Corbet, P. S., 1973, Application, feasibility, and prospects of integrated control, *in: Insects: Stud-*

ies in Population Management, (P. W. Geier, L. R. Clark, D. J. Anderson, and H. A. Nix, eds.), Ecological Society of Australia (Memoirs 1), Canberra, pp. 185–195.

Halstead, L. B., 1968, *The Pattern of Vertebrate Evolution,* Freeman, San Francisco.

Hardin, G., 1968, The tragedy of the commons, *Science* **162:**1243–1248.

Hubbert, M. K., 1969, Energy resources, *in: Resources and Man* (P. Cloud, ed.), Freeman, San Francisco, pp. 157–242.

Leopold, A., 1949, *A Sand County Almanac,* Oxford University Press, New York.

MacArthur, R. H., and Wilson, E. O., 1967, *The Theory of Island Biogeography,* Princeton University Press.

McHarg, I., 1968, Values, process and form, *in: The Fitness of Man's Environment,* Smithsonian Institution Press, Washington, D.C., pp. 209–227.

McHarg, I., 1969, *Design with Nature,* Natural History Press, New York.

Mead, M., 1970, The changing significance of food, *Am. Sci.* **58:**176–181.

Robin, E. D., 1973, The evolutionary advantages of being stupid, *Persps. Biol. Med.* **16:**369–380.

Rudd, R. L., 1971, Pesticides, *in: Environment, Resources, Pollution and Society* (W. W. Murdoch, ed.), Sinauer Associates, Stamford, pp. 279–301.

Smith, J. M., 1952, *Struthiomimus,* or the danger of being too clever (quoted by Halstead, 1968).

Southwood, T. R. E., 1972, The environmental complaint—Its cause, prognosis and treatment, *Biologist* **19:**85–94.

Strong, M., 1973, Ecologists vs. modern agriculture, *World Agr.* **22:**35–40.

Washburn, S. L., and Lancaster, C. S., 1968, The evolution of hunting, *in: Man the Hunter* (R. B. Lee and I. DeVore, eds.), Aldine, Chicago, pp. 293–303.

Watt, K. E. F., 1973, *Principles of Environmental Science,* McGraw-Hill, New York.

Wylie, P., 1968, *The Magic Animal,* Doubleday, Garden City, N.Y.

VI

The Agroecosystem: A Simplified Plant Community

Stephen Wilhelm

The production of food for man and domestic animals and of fiber is the major concern of agriculture; the managed environment that makes agriculture possible is the agroecosystem. This paper focuses on four of the pillars on which the modern agroecosystem rests: quality of management, genetical diversification of crop cultivars, industrial technology supplementing uncertainties or inadequacies of biological activity, and interactions of plant roots and soil microflora. Some of these interactions provide a high level of crop stability against soil-borne diseases and nutrient deficiencies.

The key to the term "agroecosystem" is the component *eco,* also found in ecology, economics, and ecumenical; it derives from the Greek οικος, meaning house or household. In current usage *eco* always implies the wisdom and the authority to manage, that is, decision-making and decision-following in the best interest of the household. In "agroecosystem," the idea of orderly household is expanded to include the managed environment; and the keynote of management, by and large, is best expressed by the legal phrase, "in the public interest."

The Quality of Management

The agroecosystem, that is, the managed environment, extends to wherever and includes whatever the wisdom of the learned enjoins—in practice, this

---

STEPHEN WILHELM · Department of Plant Pathology, University of California at Berkeley 94720.

is often modified by politics—and thus encompasses far more than actual farm area, farm community, patterns of land cultivation, or local farm concern. A quarantine against an alien crop pest, for instance, protects the agriculture of a district, whether or not there is any knowledge of the pest at the grower level. Similarly, local concern may be oblivious to the fact that intrusion of a new crop into the established agriculture of a particular district may jeopardize the long-standing prosperity of that district. An example is the Lindsay–Porterville olive area of the California San Joaquin Valley, where there seems to be little doubt that the encroachment of cotton cultivation into this region has triggered the recent epiphytotic losses in olives from the disease known as Verticillium wilt (Wilhelm and Taylor, 1965). Thus, an annual crop ecosystem, with its potential for mobility, if it exhibits a high degree of proneness to pest (disease) outbreaks, may jeopardize stability of the long-lived, fixed ecosystem typical of orchards and vineyards.

The agroecosystem thus has a dynamic and changing frontier. Inherently, the system is unstable and subject to upsets, and the probability of upsets increases in proportion to the diminishing of arable land. Modulating this danger today, fortunately, is agricultural research, to quote Epstein (1973), of an unparalleled scope and variety. A major objective of the research is to develop a basis for meeting the continually changing needs of crop and pest management.

In this connection it is interesting that European farmers, a few centuries ago, recognized that barberry bushes growing near their wheat fields exposed the wheat to the contagion of rust, and that after their removal, wheat was less likely to rust, or did not rust at all. However, when farmers grubbed out the bushes, especially from lands not their own, it often brought down the wrath of the townspeople, and the learned, secure in their knowledge that if the rust were a fungus, it must be the effect and not the cause of the disease, scorned the radical measures. This was so until beyond the middle of the nineteenth century. The farmer, whom science was slow to rescue from the plight of ignorance about fungi, had recognized that for wheat the agroecosystem extended beyond his fences to the hedgerows, public roadways, and into the towns, indeed, to wherever barberry grew. Thus, the agroecosystem may incidentally include a social dimension.

Nor does enlightenment necessarily ensure equanimity when diverse prerogatives, even within agriculture, seem to be in conflict. Two decades ago or more, interests arising largely from the western slopes of the California San Joaquin Valley discounted overgrazing as a factor in setting the stage for epidemic outbreaks of the curly-top disease in various important crops growing on the floor of the valley. Yet, scientific evidence showed that grazing effected changes in composition of the winter vegetation in favor of weed species such as red-stemmed filaree and plantain, which, as chance would have it, served as preferred overwintering hosts for both the virus and the insect vector. Destruction of the overwintering insect to protect the agriculture of the valley floor involved invading the western slopes with an arsenal of vehicles, chemicals and spray equipment, to say nothing of the spraying itself. The mandate of the state to impose management in the public interest, such as in places beyond areas occupied by the threatened crops, together with interrelated social, economic, and ecological problems is another aspect of the agroecosystem.

History only can validate the wisdom of the learned and separate equitable mandates from those vested in political, short-sighted, or personal interests. Idealistically, what is best for agriculture is best for all. Agriculture constitutes man's foremost and also his only essential industry (Day, 1970), and unlike other industries, it is largely a self-renewing resource: its vastly diverse commodities are produced primarily from air and water. The basic needs are fertile, deep, level soils, preferably those of valleys, pure air, pure water, and abundant sunshine, mostly conditions that now also invite urban sprawl; herein lies one of the great dilemmas of our time. "Put thy field in order," is the Biblical advice, "and afterwards build thy house" (*Prov.* 24:27). All that has been and can be built, the great city networks and their commerce, rest upon the "order of the field," that is, upon the well-managed agroecosystem. Yet, much of what is called development today by protagonists of progress disregards the old wisdom as well as the present knowledge that the farm land resource is finite and shrinking, and measures advancement only in terms of miles of freeways, sections of housing projects, and acres of concrete and asphalt parking lots. It seems that urban ecology has yet to be born.

In recent years, because of increasing public awareness of the deterioration of the environment, and, we believe, also thanks to Rachel Carson's selective documentation on pesticides in her book *Silent Spring,* decision-making in many areas of management of the agroecosystem, and particularly where pesticides are concerned, rests no longer with the agricultural community, nor even in the hands of experienced agricultural scientists. Rather, the public has become involved, and unfortunately this may carry with it an element of emotional irresponsibility. The safeguard, however, is an informed public. In our opinion, the process of agricultural education should begin in the early school years of every student of this nation and should continue at least through graduation from high school.

The Simplified Plant Community

Man is sustained by fewer than twenty species of plants. Six are the cereals: wheat, rice, rye, corn or maize, millet, and sorghum; four are root crops:

potato, sweet potato, cassava or manihot, and taro; four are legumes: bean, soybean, chick pea or Cicer, and alfalfa or lucerne; two are sugar crops: cane and beet; and two are tree crops: coconut and banana. Besides these staples, there are the beverage crops, the vine, the salad and dessert fruit crops, the oil crops, and crops grown for medicines, ornament, and perfumes. Of the fiber crops, cotton is the most important. Native to both hemispheres, it supplies many industrial raw products; in addition, cooking oil and farm animal feed are produced from the seed, which now also holds the promise of becoming even an important food source for man. No other plant provides as great a diversity of useful raw materials as cotton. Four species are cultivated, but *Gossypium hirsutum* constitutes by far the greatest acreage.

Thus, in terms of the approximately 250,000 species of angiospermous plants known, we cultivate but a very precious few. Every one is immensely important and worth any and all measures employed to protect and improve it. In the recent historical past appalling famines and severe economic depressions have resulted in several countries from diseases that devastated but single crops. For instance, Ireland lost half of its population by starvation and emigration because of the potato blight of 1845, and Ceylon was plunged into extreme depression through the coffee rust epidemic of the 1870s, a grim prospect of what may now lie in store for coffee-producing areas of Brazil. Because of the few plant species cultivated and also because of the modern practice of cultivation in pure stands, the agricultural component of the agroecosystem is actually about as simple a plant community as can be achieved. Pests among insects, mites, nematodes, viruses, fungi, bacteria, and weeds are part of the "normal," or accustomed environment of the crop, and despite management constantly put agriculture in jeopardy. Yet we cannot do as primitive cultures did and still do, except in our home gardens, namely, grow a jumble of different food plants together in small plots (Anderson, 1954), or even as the Aztecs did, plant corn and beans together, the beans climbing the corn. For the sake of efficiency of all operations and for the essentiality of high yields, modern agriculture is built around genetic uniformity, monocrop culture, and machine harvesting.

The greatest strides toward simplification, that is, the selection, and cultivation over vast areas, of intraspecies variants, now called cultivars, have been made in our day by plant breeders in the course of their exceptional success in improving crop yields.

The Diversification Principle

High-yield cultivars will continue to dominate the agroecosystem of the mechanized world, but the genetically uniform cultivar may in time yield to the genetically diverse, or multiline, cultivar. Uniform, genetically pure lines of seed-raised crops have been and still are the goals of plant breeders generally. The idea of uniformity and genetic purity is also written into the laws of crop standards. In California, for instance, industrywide planting of one pure cotton variety is mandatory by legal decree. As I write, more than one million acres are in the process of being planted to the one pure cotton strain. The strain, to ensure purity, must originate from a specific locality, namely, the United States Department of Agriculture Cotton Research Station at Shafter, California. Further, it must orginate from common Acala cotton seed stocks maintained by that station. Potentially valuable cotton cultivars developed in other states and by private seed companies thus are withheld from cotton growers of the California San Joaquin Valley. These measures, meant to be protective for an infant cotton industry 50 years ago, can scarcely be considered to be in the public interest today. Genetic uniformity of the one legal cotton variety and unhappily its unfortunate high degree of susceptibility to a virulent form of Verticillium wilt, has resulted in recent years in economic disaster for many farmers of one large, formerly very productive region of the one-variety cotton district (Figure 1).

Figure 1. Devastation by Verticillium wilt of the genetically pure cotton Acala SJ-1 exclusively planted by legal decree in the San Joaquin Valley of California. Genetical diversity holds the promise of alleviating such disease epidemics.

The disease risk, and no doubt the general pest risk, can be minimized by building specific intravarietal genetic diversity into crops. This concept applies primarily to the seed-raised crops but need not entirely exclude the vegetatively propagated crops such as potatoes, bananas, and sweet potatoes. Suneson (1960), a USDA scientist working with wheat and barley, showed two decades ago that cultivars comprising heterogeneous populations of plants resisted disease severity and spread, gave high, dependable yields year after year, and were more adaptable to diverse environmental situations than genetically pure lines. His researches pointed to many ways of building genetical diversity into crops, namely, the use of mechanical mixtures of individual lines, each with resistance to a specific fungus race, and the use of internally segregating populations in which the limits of segregation have been fairly well determined from hybridization studies of the parents. These new approaches to plant disease control through application of the diversification principle have great promise for crop improvement and allow for tailoring of the composition of cultivars to resist specific pest problems.

In our own research we are applying the diversification principle to the solution of the Verticillium wilt disease problem in cotton. We have just scratched the surface in terms of the numerous possible avenues of procedure but already the results are encouraging. For instance, if both parents of a cotton cross are resistant, the immediate offspring and the offspring of subsequent generations are also predominantly resistant. Depending on the inheritance pattern, the offspring of certain pedigreed parents are consistently superior to that of others in agronomic qualities. We believe that it is possible to combine disease resistance and standard agronomic qualities in segregating hybrid lines, which would then be used as cultivars only in one particular generation, as in F_4 or F_5. The foundation unit of seed production would always be pedigreed, cloned F_1. This system would also allow for further varying the composition of the multiline cultivar by mechanical mixing of seed of specific segregating hybrid lines. However it will be done in the future, stability of the agroecosystem in the face of established practice of monocrop cultivation will be heightened by the intragenetic diversification realized in the multiline cultivar. The excellent review of Browning and Frey (1969) is an extremely valuable contribution to this subject.

Inadequate Supply of Biological Nitrogen

Agroecosystems of developed as well as of developing countries depend on industrial chemicals, especially fertilizers. I will illustrate the point on the example of nitrogen. From ancient times, certainly from Roman times to the turn of the century, the use of manure and the cultivation of clovers, both of which supplied nitrogen, held the key to soil fertility. The Romans venerated Sterculinus, their god of the manure pile, and around 1750 the gentleman Johann Christian Schubart was knighted Edler von Kleefeld for introducing the cultivation of clover into Germany. These facts certainly indicate the great importance that was attached to soil fertility, and thus indirectly to nitrogen. Yet, the application of this knowledge could not in the long run satisfy the ever-increasing nitrogen demands of modern crops. The English chemist Sir William Crookes, in his presidential address to the British Association of Science in 1899, spoke about an impending world food shortage due to a scarcity of wheat. In 1899 the average world yield of wheat was 12 bushels per acre (16 centners per hectar). Crookes forecast one, and only one solution to the problem of world famine, namely, the transformation of atmospheric nitrogen into a form available to the plant. This the chemists subsequently accomplished, and today, 80 to 100 bushels per acre (106 to 133 centners per hectar) are common in many areas, and 200 have been achieved experimentally. Few accomplishments in science have benefited man more than development of the industrial process which supplies agriculture with synthetic nitrogen. Our daily bread depends upon it.

Clovers may again play an important role in fertilization. A better knowledge of how the symbiotic *Rhizobium* bacterium that inhabits nodules on clover roots converts atmospheric nitrogen to a form usable by the plant, and why, for instance, the bacterium does not infect wheat root tissues, may well open alternative nitrogen sources in the future.

Root Health: Prerequisite to Plant Productivity

Often a basic discovery is followed by complacency, and takes on the reputation of a panacea. This was true with the discovery of synthetic fertilizers and the traditional view of crop needs, in terms first of nitrogen, then also of phosphorus and potassium. We know now that the failure or poor yields of a crop may be due not at all to a lack of available nutrients; rather, the ability of roots to extract nutrients from the soil may have been greatly curtailed by diseases, and/or phycomycetous mycorrhizal symbionts essential to root absorption efficiency and health may be absent. Certainly root health is prerequisite to stability of the agroecosystem (Wilhelm and Nelson, 1970).

The discovery of the relation between root health and plant productivity came somewhat accidentally when, during the latter part of the 1800s, France and some other European countries were on the verge of national economic disaster due to extensive dying of grape vines from the ravages of the root-sucking aphid, *Phylloxera vastatrix*. Hundreds of thousands of acres were affected, and vine cultivation disappeared from the entire region of southeastern France. Effective relief came around 1880, in the form of carbon disulfide, some ten years after Baron Paul Thenard showed that this fumigant, injected around the base of ailing or dead vines, killed the pest and provided hope of saving the vine or of making successful replanting possible. For the first time the soil environment had been ameliorated by a synthetic, poisonous chemical, and plants other than vines, such as beets, cereals, and legumes, showed an increased growth response in the fumigated soils (Tietz, 1970).

The successful cultivation of potatoes, cotton, sugar beets, many types of orchard crops, tomatoes, and strawberries in many areas of the world today depends on soil fumigation chemicals. This dependency is the greater, the more intensive the cultivation is, that is, the more often the same crop is grown on the same land. No amount of soil fertility, manuring, clover cultivation, or even legislation could save the vine of the past century from *Phylloxera,* nor would they protect the crops of today against the myriad of known root-destroying pests. The only rapid and efficient control known today is offered by the volatile chemical.

In California, but also elsewhere, chloropicrin and methyl bromide, because of their exceptional properties in controlling soil-borne diseases, are now widely used for soil fumigation where intensive agriculture is practiced; being synergistic, they are applied simultaneously (Figure 2). By ridding the soil environment of pests among insects, fungi, nematodes, and weeds, they promote healthy root growth (Figure 3), and exceptional crop production. Because of their remarkable salutary properties, which well offset the risks connected with their manufacture and use, these chemicals appear to act as powerful fertilizers, but fertilizer alone would never correct the soil problems they do.

The consistent salutary effects resulting from proper use of soil fumigants—I am most familiar with methyl bromide–chloropicrin mixtures—suggest that not only are injurious soil-borne organisms controlled, but that beneficial ones are favored. Beneficial species have been identified among the soil bacteria of the genus *Bacillus*. Occupying the rhizosphere, they may protect plant roots from infection by fungi or suppress pathogen activity in other ways. They may favor crop growth by producing growth-stimulating hormones such as indolacetic acid or giberellic acid, by mineralization of the soil, or by destruction of toxins of microbial origin that injure crops (Broadbent *et al.*, 1971). The complex interactions of these beneficial species among themselves, with other components of the soil microflora and microfauna, and with the cultivated plant, are essential to the stability of the agroecosystem. Careful

Figure 2. Intensive agriculture requires commercial soil fumigation for control of soil-borne pests. Here a mixture of methyl bromide and chloropicrin is being applied prior to planting of strawberries. Simultaneously with injection of the chemicals the soil is covered with polyethylene sheeting.

Figure 3. Right, exceptional strawberry root growth, achieved experimentally 25 years ago, in soil freed of pests by fumigation with chloropicrin. Left, root system of the "normal" plant.

management of the ecology of the plant root, particularly of the rhizosphere biology, thus will be a critical aspect of agroecosystem management in the future.

Rhizosphere management involves research on many facets of soil and root biology. One area now of foremost importance is the endophytic phycomycetous mycorrhizal system involving the fungus *Endogone*. *Endogone* has been observed for more than 75 years in roots of a great variety of cultivated and noncultivated plants. But as attempts to isolate the fungus and grow it in the laboratory have not yet been successful, little more has come of it than that all who ever viewed the extensive inter- and intracellular parasitism wondered about its significance. To cite Gerdemann (1968), *Endogone* parasitizes more angiospermous plant tissue than all other parasites, including pathogenic infections, put together. In exchange for the parasitism, as it were, the *Endogone* fungus absorbs nutrients from the soil and translocates them to the host plant. It is best known for its ability to absorb phosphorus, but it probably absorbs other elements as well, even micronutrients and water, beyond what is otherwise available to the plant. Its extensive, extracellular mycelium appears at times to take the place of root hairs, which are lacking in most crop plants, especially perennials. Another remarkable fact is that root tissues occupied by *Endogone* may be protected against invasion by injurious fungus parasites. Thus, *Endogone* may be a major chthonic factor in the biological control of soil-borne, root-infecting plant diseases.

Healthy *Endogone,* that is, the health of both hyphae occupying the soil, and hyphae and complex arbuscules occupying living root tissue, thus is prerequisite to crop health. Manipulation of the *Endogone*-mycorrhizal system, therefore, seems to offer unparalleled opportunities for contributing to agroecosystem stability. With further research, no doubt, host-preferring strains of *Endogone* will be found, as well as strains with different degrees of efficiency in capacity to absorb nutrients. Also, specific strains may be employed in the biological protection of roots from pathogen attack, and techniques will eventually be developed whereby crops may be inoculated with specific strains of *Endogone* at the time of planting. Interactions between *Endogone* and rhizosphere microorganisms have still to be explored.

Though we view the agroecosystem as a simplified plant community, which it is in terms of cultivation practices, it is supported below ground by microbial interactions that almost defy unraveling, and it is microbial complexity that effects stability. Roots, being out of sight, unfortunately have not been in the spotlight of agricultural research to the extent that they deserve; yet upon their health, the quality of bacterial rhizosphere inhabitants, and on internal fungal symbionts, depends the productivity of the crops of all agroecosystems.

Literature Cited

Anderson, E., 1954, *Plants, Man and Life,* A. Melrose, London. 208 p.

Broadbent, P., Baker, K. F., and Waterworth, Y., 1971, Bacteria and actinomycetes antagonistic to fungal root pathogens in Australian soils, *Austral. J. Biol. Sci.* **24:**925–944.

Browning, J. A., and Frey, K. J., 1969, Multiline cultivars as a means of disease control, *Ann. Rev. Phytopathol.* **7:**355–382.

Day, B. E., 1970, Operation eco-perspective, *Calif. Agr.* **24(10):**2.

Epstein, E., 1973, Agriculture, research, and shortages of funds and food, *Science* **181:**997.

Gerdemann, J. W., 1968, Vesicular-arbuscular mycorrhiza and plant growth, *Ann Rev. Phytopathol.* **6:**397–418.

Suneson, C. A., 1960. Genetic diversity—A protection against plant diseases and insects, *Agron. J.* **52:**319–321.

Tietz, H., 1970, One centennium of soil fumigation: Its first years, *in: Root Diseases and Soil-Borne Pathogens* (T. A. Toussoun, R. V. Bega, and P. E. Nelson, eds.), University of California Press, Berkeley, pp. 203–207.

Wilhelm, S., and Taylor, J. B., 1965, Control of Verticillium wilt of olive through natural recovery and resistance, *Phytopathology* **55:**310–316.

Wilhelm, S., and Nelson, P. E., 1970, A concept of rootlet health of strawberries in pathogen-free field soil achieved by fumigation, *in: Root Diseases and Soil-Borne Pathogens* (T. A. Toussoun, R. V. Bega, and P. E. Nelson, eds.), University of California Press, Berkeley, pp. 208–215.

VII

Tobacco Pest Management

R. L. Rabb, F. A. Todd, and H. C. Ellis

Introduction

Tobacco is a solanaceous plant which originated in the Americas; but, since colonial days, it has spread throughout the world. Since Sir Walter Raleigh's day, the tobacco plant and its culture have been altered dramatically, and changes wrought by plant breeders and other technologists have been particularly rapid during the past 30 years. Today, there are many classes, types, and varieties of tobacco grown under many different environmental conditions; and, in the different ecosystems modified for tobacco culture, there are many pest species among the associated flora and fauna.

This paper is restricted largely to flue-cured tobacco and its major disease and insect pests in North Carolina. The term "pathologist" is used in a broad sense to denote expertise in dealing with diseases caused by viruses, bacteria, and fungi as well as nematodes.

To place our subject in a spatial perspective, we call attention to the geographic distribution of tobacco in the United States. The six major classes of tobacco (flue-cured, fire-cured, air-cured, cigar-filler, cigar binder, and cigar-wrapper) are largely limited to the East Coast, and the flue-cured type makes up over 50% of the crop grown in the United States. North Carolina ranks first among the states in its production.

Tobacco production in North Carolina has evolved with a relatively high

---

R. L. RABB · Department of Entomology, North Carolina State University, Raleigh, North Carolina 27607. F. A. TODD · Department of Plant Pathology, North Carolina State University, Raleigh, North Carolina 27607. H. C. ELLIS · Department of Entomology, North Carolina State University, Raleigh, North Carolina 27607. Current address: Georgia Coastal Plain Experiment Station, Tifton, Georgia 31794.

degree of success as a result of disciplinary and interdisciplinary research and extension efforts. Systems science and particularly advanced techniques of systems analysis are "in" now and much progress may follow. However, we should not lose sight of the fact that interdisciplinary effort by specialists well grounded in basic and applied disciplines is the crucial requirement for successful systems science. To date, our systems approach entails largely mental and pictorial constructs, and we have only just begun to use comprehensive mathematical models and advanced techniques of systems science, with concentration of these efforts on certain subsystems where black boxes seem to be of particular importance to practical goals.

In our mental construct of the tobacco ecosystem, populations of various species of tobacco pests are conceived as subsystems interacting with other subsystems of a seminatural system maintained in its seminatural state by cultural practices, inputs of energy subsidies. The central element in this system, and that to which all other subsystems are subservient from man's view, is the tobacco plant. Thus, to understand tobacco pest management one must first understand tobacco, whose biological and ecological characteristics and essential cultural practices set certain *a priori* limitations on the options for managing pest subsystems. The essential features of flue-cured tobacco production of most relevance to pest management are as follows.

Tobacco seeds are sown in beds from January to mid-March and protected during growth by cloth or plastic coverings. Seedlings are pulled from these beds in April and May and set in fields, an operation known as *transplanting*. The main season for plant growth is in June and July, and during the earlier stages of growth, two or more cultivations are made. When flowering occurs, the inflorescences are broken off and discarded, a process known as *topping*. Topping channels more nutrients into marketable leaves, but this is a mixed blessing since it forces the growth of lateral buds in the leaf axils. These buds, called *suckers,* produce considerable foliage and flowers of no commercial value; consequently, they must be controlled for maximum production of marketable leaves.

We shall give major attention to the control of suckers because of their influence on pest population dynamics (see Seltmann, 1971, for review of modern methods of tobacco sucker control). Prior to the late 1950s, suckers were removed by hand, a practice that was repeated several times because the five to six primordial buds in each leaf axil produce successive "new" suckers as old ones are removed. In the late 1950s, chemical methods of sucker control were developed, and the principal chemical used was maleic hydrazide (MH). When sprayed on plants after topping, MH is absorbed systemically and inhibits cell multiplication, but not cell growth. Consequently, MH properly used will eliminate sucker growth during most of the leaf maturation and ripening period.

The tobacco leaves ripen from the bottom to the top of the stalk and are harvested in four to six primings at 6–10 day intervals. The harvested leaves are dried (cured) in various types of artificially heated barns. The older barns were equipped with wood-burning furnaces and metal flues—hence, the name flue-cured—but other fuels, chiefly oil, and more modern curing processes are currently used.

The timing of production practices from seeding through harvest are reasonably predictable and uniform throughout North Carolina, with proper allowance for seasonal differences from the mountains to the coast; however, there is great variation in the timing and nature of post harvest practices. Stalks are left standing in some fields for various lengths of time ranging from a few days to several months. In some cases, stalks are merely cut and allowed to produce regrowth from "ground suckers"; whereas, in other cases, stalks are cut and roots are plowed out. Most growers destroy stalks before frost, but timing and method of destruction is variable.

Disease and Insect Control Subsystems

Some Basic Considerations

Disease and insect control are integrated into the tobacco system chronologically from preplanting to postharvest as shown in Table I. Ten steps of

TABLE I. Chronology of Pest Management Inputs into Tobacco Production

Steps of tobacco production chronologically	Pest control disciplines principally involved at each step
1. Field selection (rotation)	Pathology
2. Variety selection	Pathology
3. Plant bed practices	Pathology and entomology
4. Plowing, disking, fertilization (preplanting)	
Herbicides	Weed science
Nematicides	Pathology
Multipurpose pesticides	Pathology and entomology
Soil insecticides	Entomology
5. Planting	Entomology (secondarily)
6. Inputs during plant growth	
Cultivation, top-dressing, irrigation	Weed Science
Foliar pesticides	Entomology
7. Topping and sucker control	Entomology (secondarily)
8. Harvest 4–6 primings	Entomology (secondarily)
9. Stalk destruction	Entomology and pathology
10. Plowing	Pathology and entomology

tobacco production are listed, and the pest control disciplines principally involved are indicated for each step.

The pathologist dominates the decisions regarding field selection on the basis of disease problem history. He also is the principal decision-maker in varietal selection as prompted by the disease history of the field site. Pathologists, entomologists, and weed scientists contribute to decisions on preplant soil treatments. Entomologists make *ad hoc* insect control decisions during the period from planting until harvest, and both pathologists and entomologists make important recommendations regarding crop residue disposal.

Integrating inputs of various specialists is not always easy, and both positive and negative interactions are encountered. This is illustrated by the interactions involved among inputs for hornworm, sucker, and nematode control.

A few key facts concerning the tobacco hornworm, *Manduca sexta* (Linnaeus) (Figure 1) are essential to an understanding of the interrelationships to be discussed.

The adult moths are strong-flying, nocturnal, and nectar-feeding. Females oviposit preferentially on succulent leaves of a number of solanaceous plants. In major tobacco-producing areas of the United States, tobacco represents the

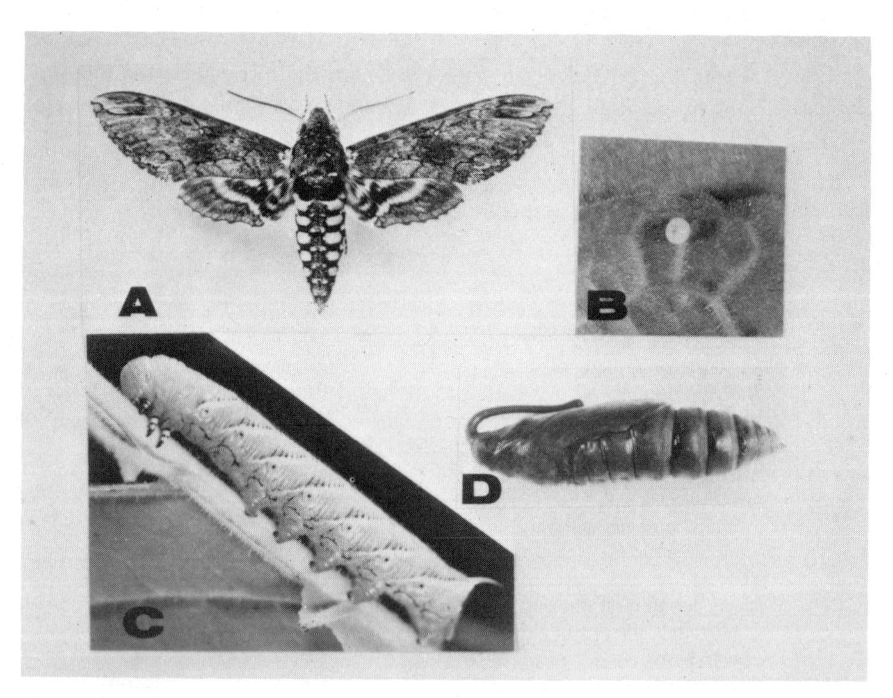

Figure 1. Life cycle of the tobacco hornworm: A, adult; B, egg; C, larva; D, pupa.

Figure 2. Tobacco stems with foliage stripped by feeding of tobacco hornworms.

principal biomass of host plant tissue consumed by the larvae hatching from these eggs. These developing larvae are capable of defoliating succulent plants (Figure 2), and over 75% of the damage is inflicted in the last of the five larval instars. Fully developed larvae enter the soil and pupate at a depth of 5–8 in. (See Madden and Chamberlin, 1945, as one of many references to tobacco hornworm biology.)

The seasonal history of the hornworm is intimately linked to the phenology of tobacco, and its populations are strongly influenced by the sequence of tobacco production practices (Figure 3). It overwinters in the pupal stage, and emergence begins in May and extends into July. The first two generations are of greatest concern to tobacco growers, since they attack marketable tobacco, but the third and partial fourth generations, which occur primarily after harvest on noncommercial sucker growth, are of most significance to populations of the succeeding year. The majority of the pupae of these last two generations enter

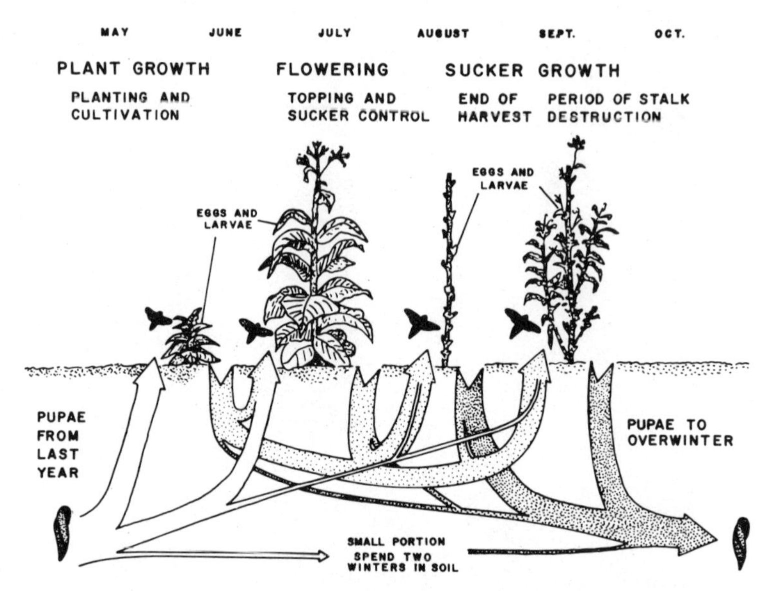

Figure 3. Seasonal history of tobacco hornworms as associated with the seasonal stages of tobacco production in North Carolina.

diapause, a dormancy state initiated during the larval period by photoperiod (Rabb, 1966). Very few individuals enter diapause before the second week in August, but then the incidence increases abruptly to over 95% of all pupae produced. Only those in diapause have the potential to overwinter successfully in areas with cold winters.

From the viewpoint of developing management procedures for hornworm populations, the diapause characteristics of the species and the environmental factors influencing the production and survival of diapausing pupae are of central importance.

Due to innate characteristics and phenology of cultivated tobacco, diapausing pupae are produced largely on tobacco foilage (sucker growth) of no commercial value in fields where stalks are left standing after harvest. Destruction of these stalks immediately after harvest has been recommended for many years as a means of eliminating food for potentially overwintering insects (Metcalf, 1909; Rabb, 1969) as well as for disease and nematode control (Nusbaum, 1959; Nusbaum and Todd, 1970; Todd, 1971*a*). Although many farmers have followed these recommendations, compliance has not been uniform. The difference between the realized and potential reduction in sucker growth after harvest is indicated in Figure 4, which is a generalized graph depicting the rise and fall of green tobacco on an acreage basis in North Carolina. Harvesting and stalk destruction proceed rapidly in August and result in a sharp decline in the acreage of available oviposition sites for hornworm adults. The cross-hatched area in the graph represents stalks with varying degrees of sucker growth that could have been eliminated by prompt stalk destruction immediately after harvest.

While stalk destruction has been and continues to be one of the most important factors influencing the production of diapausing hornworms, another factor was introduced into the tobacco production system in the late 1950s which reduced, *but did not eliminate,* the significance of postharvest standing stalks. This factor was the advent of chemical sucker control. As noted earlier, MH is a widely used systemic growth regulator that inhibits sucker growth not only from the time of application through harvest, but continues to retard the development of suckers long after harvest; in some cases, completely. By the early 1960s, MH was in general use throughout the entire flue-cured tobacco belt, and growers experienced an indirect benefit in the form of a general decline in hornworm populations because of the decreased food supply for overwintering hornworms (Rabb *et al.* 1964; Rabb, 1969).

Thus, the interaction between hornworm control and sucker control has been positive, with a significant benefit in terms of reduced hornworm outbreaks resulting from a practice developed for another purpose. Tobacco production, however, is dynamic and new inputs have influenced both sucker production and hornworm populations. One such input has been the release of tobacco varieties resistant to nematodes (Clayton *et al.*, 1958; Moore *et al.*, 1962).

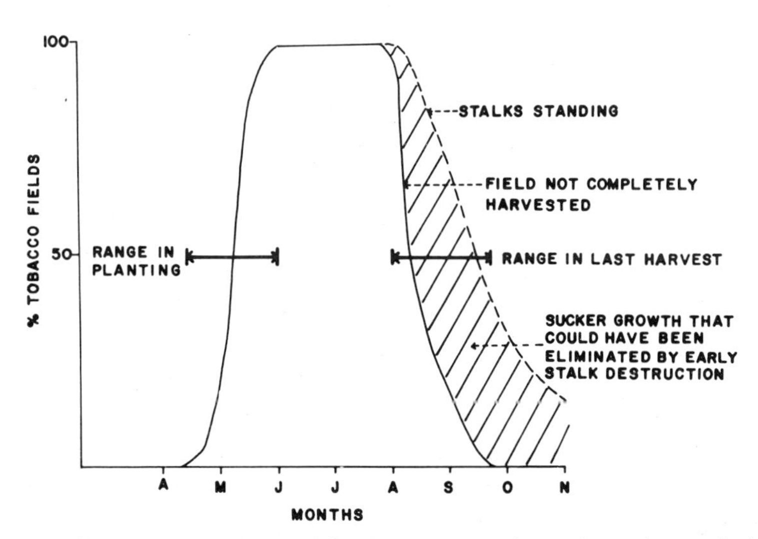

Figure 4. The seasonal rise and fall in acreage of growing tobacco in North Carolina (somewhat diagrammatic).

The first indication of a change in the system was an increase in the number of postharvest fields with noticeable sucker growth. Even though most of this increased sucker growth was malformed due to residual effects of earlier applications of MH, it was of potential significance to postharvest insect production. One factor involved in the increased sucker production was the introduction of the nematode-resistant varieties which have more vigorous root systems, especially in late summer and fall after harvest, than nematode-susceptible varieties that are damaged by nematodes and often predisposed to infection by root disease organisms. Consequently, even with the earlier applications of MH, the root stocks of nematode-resistant plants produce more postharvest sucker growth than susceptible varieties. Sucker growth on resistant and susceptible varieties grown in plots with and without nematodes is contrasted in Table II. These data indicate approximately a threefold increase in postharvest sucker production where nematode resistant varieties are grown in nematode-infested soil.

The changing relationships between hornworm populations, sucker control, and nematode-resistant varieties are summarized in Table III. Since the introduction of MH, hornworm populations have been low and have remained low even after some increase in postharvest sucker growth due to nematode-resistant varieties. However, our information is too scanty for definite conclusions regarding cause and effect. An increase in the number of fields with early stalk destruction may be partly responsible for holding the hornworm populations in check in spite of greater sucker production potential. Another important factor is the density-related action of natural enemies, particularly *Apanteles congregatus* (Say) and *Winthemia manducae* (Sabrosky and DeLoach). The abrupt reduction in oviposition sites for hornworm females brought about by sucker control and stalk destruction results in the concentration of hornworm eggs after harvest on a small fraction of the preharvest tobacco acreage. This

TABLE II. Postharvest Sucker Weights[a] of Tobacco Varieties Resistant and Susceptible to Nematodes (*Meloidogyne incognita*), 1973.
(Data supplied by T. E. Reagan, North Carolina State University)

Soil treatment	Tobacco variety: NC 2326 (Susceptible)	Tobacco variety: PD 79 (Resistant)
Inoculated with nematodes	0.67	2.10
Fumigated with methyl bromide	2.42	2.42

[a] Number = pounds of suckers per 10 plants (3 replications).

TABLE III. Simplified Representation of Changing Relationships Between Hornworm Populations, Sucker Control Practices and Nematode Resistant Varieties

Chronology	Type of sucker control	% Acreage nematode-resistant varieties	Postharvest sucker growth	General hornworm problem
Prior to 1960	Manual	0	Heavy	Serious
1960–1969	Maleic hydrazide (MH)	0.5	Very light (Malformed)	Minor
1969–present	MH plus new materials	50	Light but variable (Mostly malformed)	Minor

quickly reduced acreage also reduces the host plant area in which parasites search for hornworms. Thus, this man-induced concentration of the general hornworm population into the relatively few fields producing succulent sucker growth sets the stage for very heavy hornworm mortality due to parasitism and intraspecific competition.

This is but one example of interactions among the principal elements of the tobacco ecosystem and among the inputs of tobacco production specialists. Other examples include the role of insects in the transmission of disease (Bradley and Ganong, 1957), the increased susceptibility of plants to disease because of insect damage (Nusbaum *et al.*, 1961), and the influence of MH on the incidence of brown spot disease caused by *Alternaria longipes* (Von Ramm *et al.*, 1962). MH may also inhibit or retard development of *Meloidogyne* spp. (root-knot nematode) and galls on tobacco (Nusbaum, 1958; Davide and Triantaphyllou, 1968). Thus, there is ample justification for increased interdisciplinary effort in research and extension.

Action Programs in Tobacco Pest Management

The integration of pest control actions into the tobacco production system has been a slow step-by-step process taking many years. Most of the inputs by pathologists and entomologists have been made separately, and each group has developed and implemented its own "action" program. Each shall be described briefly.

Management of Insect Pests

A pilot insect pest management program, funded through the Animal and Plant Health Inspection Service (APHIS) of the U.S. Department of Agriculture, was initiated in 1971. It represented a cooperative effort of state and federal personnel (North Carolina Agricultural Extension Service, North Carolina Agricultural Experimental Station, North Carolina Department of Agriculture, USDA-APHIS). The following brief account has been abstracted from the Annual Reports (Gaynard *et al.*, 1971; Ellis *et al.*, 1972; and Ellis and Gaynard, 1973).

The program was aimed primarily at the two major foliar insect pests of tobacco: the tobacco hornworm, previously discussed, and the tobacco budworm, *Heliothis virescens* (F.), whose life cycle also includes adult moth, egg, larval, and pupal stages and whose seasonal history involves three to four generations each year (Figure 5). Damage is inflicted by the larvae feeding in the terminal vegetative buds of growing tobacco (Figure 6), and the problem period essentially ends with the topping of tobacco, particularly if sucker control is effective (Reagan *et al.*, 1974).

Personnel of the insect pest management action program had a relatively rich source of information as a point of departure, including accounts of both successes and failures in control. Since the turn of this century and particularly since World War II, many different insecticides were used on tobacco, and some of these caused problems. Certain materials proved to be a source of off-

Figure 5. Life cycle of the tobacco budworm: A, adult; B, egg; C, larva; D, pupa.

Figure 6. Tobacco budworm feeding in the vegetative terminal of tobacco.

flavor (Allen *et al.,* 1952; North Carolina State University Entomology Faculty, 1958); consequently, flavor tests became standard procedure in screening new materials for use on tobacco. Insecticidal residues as a potential health problem also received and continue to receive careful study (Weber, 1956; Guthrie *et al.,* 1959*a*; Bowery *et al.,* 1959, 1965; Lawson *et al.,* 1964; Guthrie, 1968, 1973; Sheets *et al.,* 1968; Domanski *et al.,* 1973; Tappan *et al.,* 1973). Some cases of insecticidal resistance were encountered (Guthrie *et al.,* 1963; Rabb and Guthrie, 1964). Chlorinated hydrocarbons largely were eliminated from use on tobacco, and now certain carbamates, organic phosphates, and the microbial *Bacillus thuringiensis* are in use (Rabb and Guthrie, 1956; Rabb *et al.,* 1957; Guthrie *et al.,* 1959*b*; Allen *et al.,* 1961; Creighton, *et al.,* 1961; Dominick, 1968; Mistric and Smith, 1973*a,b*). Researchers also developed practical economic thresholds for hornworms and budworms, and methods of chemical application were devised to reduce deleterious effects on natural enemies (Guthrie *et al.,* 1959*c;* Lawson and Rabb, 1964; Mistric and Smith, 1969; Mistric and Pittard, 1973; Elsey, 1973*a*). Biological studies of insect pests and some of their important natural enemies in North Carolina, especially of hornworms and budworms, were available (Fulton, 1940; Lawson, 1959; Rabb and Lawson, 1957; Rabb, 1960, 1966, 1971; Rabb and Bradley, 1968; Neunzig, 1969; Rabb and Thurston, 1969; DeLoach and Rabb, 1971, 1972; Elsey and Stinner, 1971; Elsey, 1972, 1973*b;* McNeil and Rabb, 1973*a,b*). Investigations of the impact of sucker control and stalk destruction (Rabb *et al.*, 1964; Kinard *et al.*, 1972) gave a sound basis for cultural control, and guidelines for a tobacco insect pest management program had been proposed (Lawson *et al.*, 1961; Rabb, 1969).

In spite of the available expertise and technology, however, various assessments of actual grower practices indicated great potential for improvement. Thus, the cooperative action program was launched in 1971 with five principal objectives, two operational strategies, and three principal management practices, as shown in Figure 7. These objectives, strategies and practices, with the exception of the sucker control practice, are common to many agricultural pest control programs. Of the five objectives, prevention of undesirable residues is of particular significance in tobacco production, because the physical and chemical characteristics of the leaf are conducive to retention of residues of many types of chemicals, and the tobacco market is very sensitive to foreign residues of any kind.

The last two objectives, i.e., the stabilization of pest populations at low levels and the maintenance of high economic thresholds, are closely dependent on the degree to which all practices are integrated with natural enemy action. This is illustrated by the integration of insecticidal and cultural control tactics with the actions of four enemies of hornworms: *Jalysus spinosus* (Say), *Polistes* spp., *Apanteles congregatus* (Say), and *Winthemia manducae* (Sabrosky and DeLoach) (Figure 8).

As shown in Figure 9, *J. spinosus* and *Polistes* spp. are more active during the preharvest period, when they reduce the numbers of hornworms surviving to the fifth instar. Economic damage (i.e., damage exceeding cost of prevention) has not been authenticated prior to the appearance of these large fifth in-

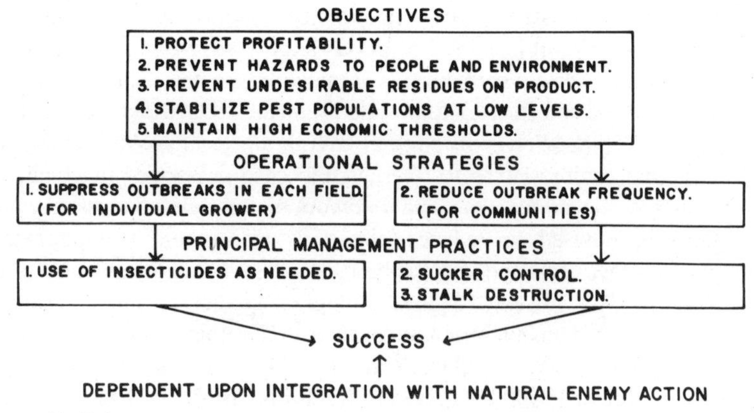

Figure 7. Principal objectives, operational strategies, and management practices of tobacco insect pest management.

Figure 8. Four principal natural enemies of the tobacco hornworm in North Carolina: A, *Jalysus spinosus;* B, *Polistes* sp.; C, *Apanteles congregatus* cocoons on back of hornworm larva; D, *Winthemia manducae* adult near hornworm larva.

stars, and measurements indicate that 85–95% of hornworm damage is inflicted by them. On the other hand, the number of hornworm eggs and small larvae is a poor indicator of the number of fifth instars to be expected because of the actions of natural control factors, such as *J. spinosus* and *Polistes* spp. (Lawson and Rabb, 1964). These natural control agents do not always suppress hornworm populations below the level of economic loss, but they are very effective in a majority of fields during most seasons. By basing the economic threshold on the number of large larvae (large fourth and fifth instars) per unit area at the time first fifth instars appear, the grower can take maximum advantage of the direct impact of the natural enemies in reducing leaf loss on his current crop (Lawson *et al.*, 1961). From a population dynamics view, these two enemies reduce the number of hornworm pupae produced; and most, but not all, of them are nondiapausing because the mortality inflicted occurs primarily prior to diapause initiation in the general hornworm population. However, the other two enemies, *A. congregatus* and *W. munducae*, are more important in postharvest tobacco fields and reduce the number of overwintering hornworms and next

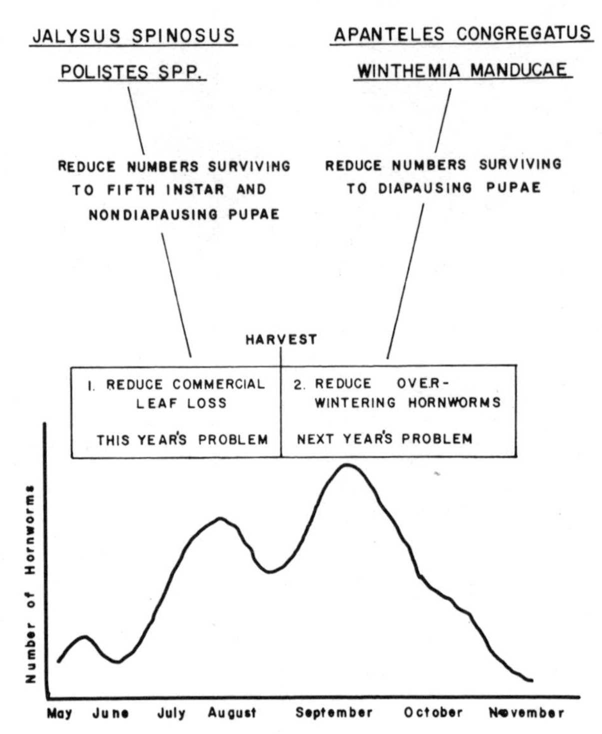

Figure 9. Two important actions of four tobacco hornworm enemies (simplified and generalized).

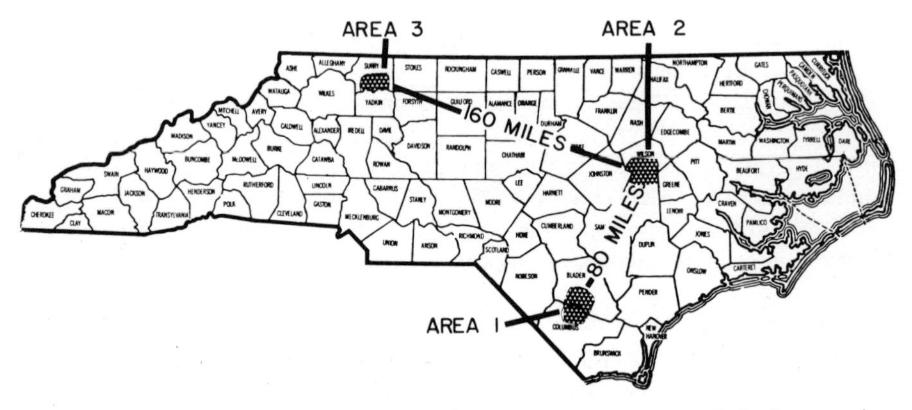

Figure 10. Location of three tobacco insect pest management study areas in North Carolina.

year's problem. There is no negative interaction between these parasites and insecticides after harvest and parasite effectiveness has not been reduced by the effects of MH sucker control or by stalk destruction.

The tobacco insect pest management program varied somewhat in size each year, but in 1973 involved approximately 813 farms and 11,350 acres of tobacco in three study areas as shown in Figure 10. Area 1 is in the Southeastern Coastal Plain, area 2 is in the Central Coastal Plain, and area 3 is in the Northwestern Piedmont. Thirty-three full-time summer employees scouted over 3400 tobacco fields each week during the growing season. Among the data collected were counts of certain pests and beneficial insects, plant growth and phenological events, cultural practices, pesticide applications, insect damage, and phenology of surrounding crops.

Data processing was complex, as shown in Figure 11, but it was handled so that growers were informed immediately when pest levels reached the following thresholds: budworms, 5 or more plants per 50 plants having living larvae prior to buttoning (the appearance of flower buds in plant terminals); hornworms, 5 or more larvae 1 in. long or longer (fourth and fifth instars) per 50 plants; flea beetles or aphids, when infestations were rated "heavy" according

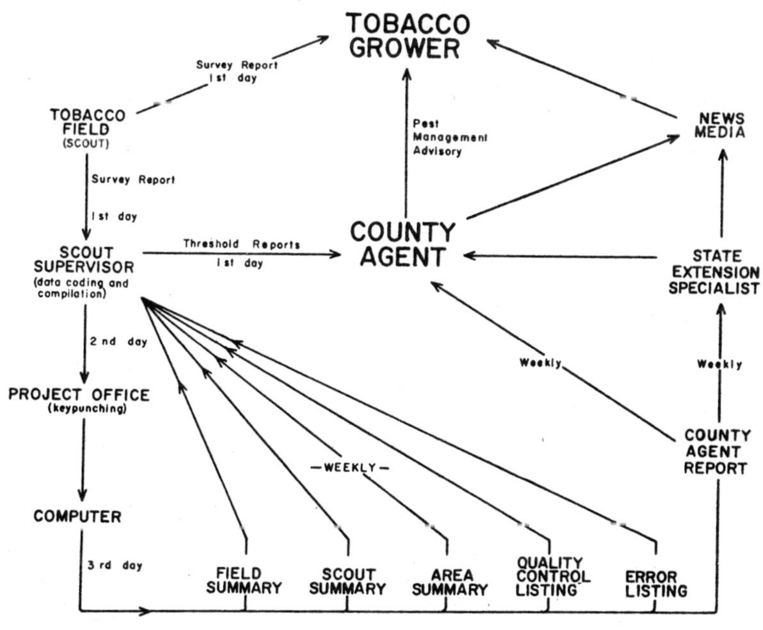

Figure 11. Data flow system used in the tobacco insect pest management program.

to an objective rating procedure. All data were stored on computer tapes for summary and retrieval.

Population levels for hornworms, budworms, and for four of the beneficial entomophagous insects, as determined by scouting reports in 1973, are depicted in Figure 12. Hornworm levels followed the same trend as detected in 1971 and 1972, being much heavier in area 1, intermediate in area 2, and lowest in area

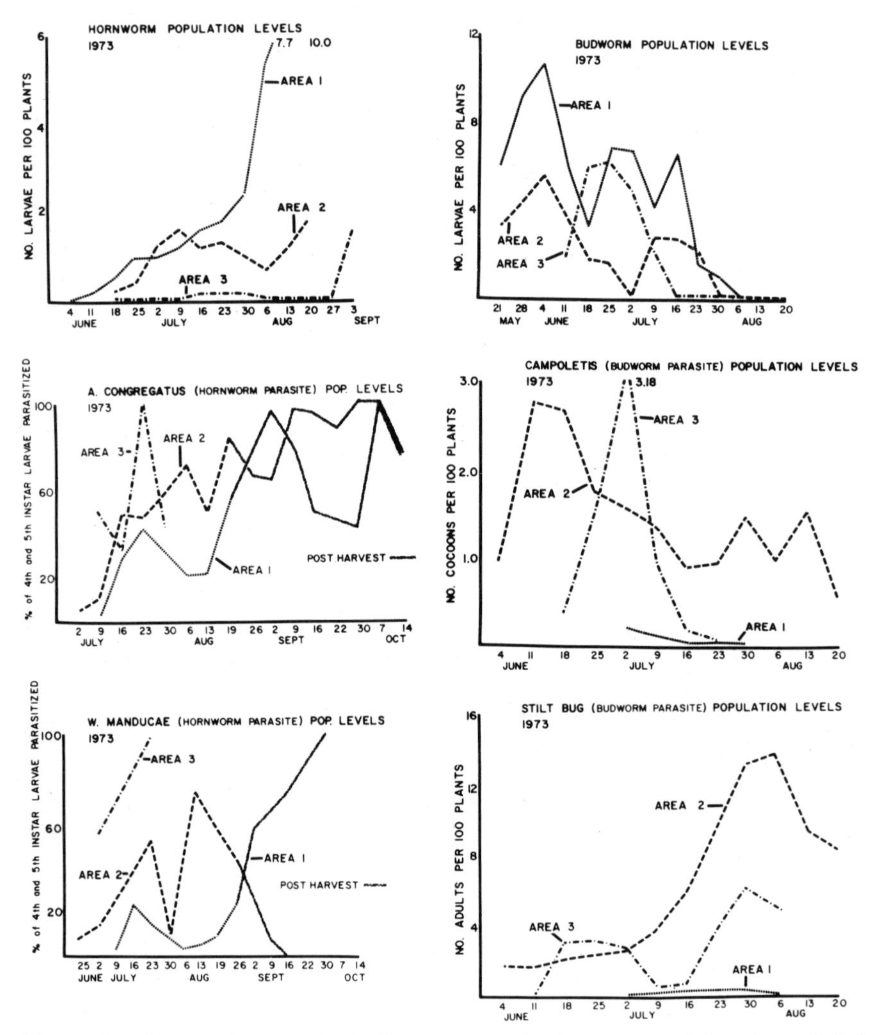

Figure 12. Population levels for hornworms, budworms, and four beneficial insects in three North Carolina study areas in 1973.

3. The same trend has been detected for budworms in 1971 and 1972, but 1973 was an outbreak year for *Heliothis* (both the tobacco budworm and corn earworm, the latter also infesting tobacco), and differences in budworm levels between areas were less clearcut, due not only to the heavier general populations but also to heavier and more widespread use of insecticides for their control.

The three years' data indicated that populations of beneficial insects, particularly prior to harvest, were lowest in area 1, highest in area 3, and intermediate in area 2. The 1973 data followed this pattern except that counts of budworm enemies were higher in areas 2 and 3. *Campoletis sonorensis* (Cameron) populations were higher in both areas, and stilt bugs were more numerous in area 2 than in 1971 and 1972. Stilt bugs are important predators of eggs and small larvae of *Heliothis* spp. as well as of hornworms, and *Campoletis* is a principal larval parasite of *Heliothis* spp. in North Carolina. Populations of these two enemies were monitored not only because of their importance but also because monitoring posed less problems than for most other enemies. One might have expected increased populations of budworm enemies in response to the higher budworm populations in 1973. However, there was little indication of increases in *Campoletis* or stilt bugs in area 1, where budworm populations were consistently high and where insecticidal treatments have been heavier than in other areas. There was also a decrease in stilt bug levels in area 3 during 1973 accompanying the increase in budworm populations. Circumstantial evidence suggests that the reduction in stilt bugs in area 3 may have been related to the increased use of insecticides triggered by the general budworm outbreak. The percentage increase in treatments was greater in area 3 than in other areas. Some increase in hornworm parasitization in 1973 was noted, but the trend for populations of parasites to be lowest in area 1 and highest in area 3 held.

The high budworm population levels in 1973 produced approximately a threefold increase in the number of threshold reports (i.e., the number of counts of insect pests that equaled or exceeded the economic threshold). This increase was fourfold in area 3. On the other hand, hornworm populations were slightly lower in all three areas in 1973 and so were the number of threshold reports for hornworms, except in one county in area 1. Insecticidal use increased approximately 6% in 1973, largely as a reaction (in some cases an overreaction) to the budworm outbreak. However, the increase was not proportional to the increase in threshold reports. Since insecticidal use was excessive in previous years, the modest increase in treatment in 1973, in spite of heavier budworm populations, might be due to increased use of the threshold principle in decisions regarding insecticidal use. Theoretically, there should be no more than one application per threshold report. The actual numbers of applications per threshold report in counties of the three areas during the three years of the program are given in Table IV. In four of the five counties involved, there was steady progress in reducing unnecessary insecticidal use, as indicated by the closer approximation

TABLE IV. Excessive Insecticide Use as Reflected by the Number of Foliar Applications Per Threshold Report in 1971, 1972, and 1973

		Average number of foliar insecticide applications per threshold report [a]		
Area	County	1971	1972	1973
I	Columbus	2.5	3.7	1.3
	Bladen	2.5	1.9	1.2
II	Wayne	5.3	2.3	1.1
	Wilson	5.3	3.5	4.8
III	Surry	4.0	2.0	0.7
	TOTAL	3.4	2.6	1.6

[a] Ideal is a figure of 1.0, one application per threshold report.

of application numbers and threshold events from 1971 through 1973. Although the average number of applications per field increased from 1.2 in 1972 to 1.9 in 1973, the excessive use of insecticides decreased in 1973 by 38.5% as compared with 1972 and 52.9% as compared with 1971. The trend of the data in Table IV indicates progress toward the goal of optimum use of insecticides on tobacco. However, there is still need for improvement in both reducing excessive numbers of applications and in the proper timing and methodology of treatment. Some growers continue to apply insecticides almost weekly without regard to population levels, and in some cases application rates and methods are improper.

In addition to promoting the use of the economic threshold principle by individual growers to suppress outbreaks in each field, the personnel of the program and other tobacco specialists continued to recommend effective sucker control and early stalk destruction as a means to reduce winter carryover of pests and outbreak frequency. Postharvest surveys were made to determine the status of stalk destruction, and there was considerable variation from county to county in efficiency (Figure 13). Although there was noted improvement in 1972 as compared with 1971, little or no increase in stalk cutting *per se* was detected in 1973. Improvement, however, was noted in the timing and method of stalk destruction and in the number of fields in which stalks were plowed out, disked, and sowed to winter cover crops. These latter improvements are of great significance to disease and erosion prevention.

Figure 14 illustrates the estimated production index for nonparasitized fifth-instar hornworms diapausing in fields with standing stalks in the Columbus–Bladen and Wayne–Wilson areas in 1973. It also illustrates the potential for reduced diapausing hornworms by appropriately timed stalk destruction. To determine the weekly production index, the average number of nonparasi-

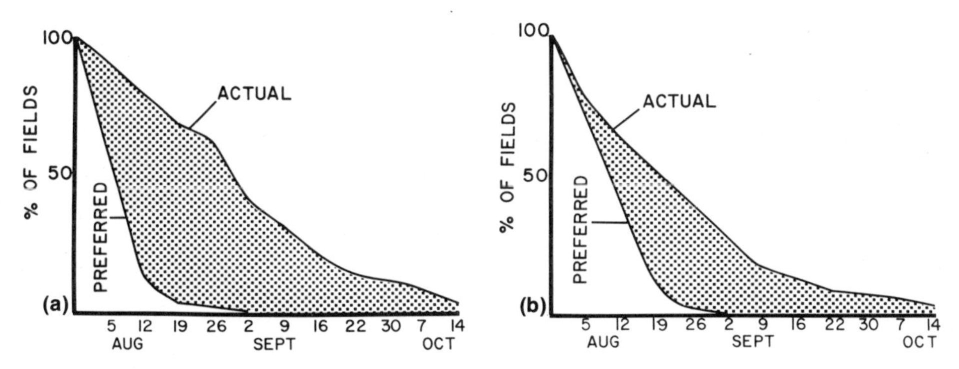

Figure 13. Time lag between preferred and actual stalk destruction in Bladen (a) and Wilson (b) Counties in North Carolina, 1973.

tized, fifth-stage larvae diapausing per 100 plants per week was established on the basis of weekly insect counts and the average percent of larvae entering diapause per calendar date as determined by Rabb (1966). This average, when multiplied by the number of standing fields per weekly interval provided the production index figure. It was assumed that all nonparasitized larvae pupated successfully.

In the Columbus–Bladen area in 1973, the period of highest production for overwintering hornworms was again during August, with peak period occurring about one week earlier than in 1972. The production during September, 1973, in the Columbus–Bladen area was greater than in 1972, primarily because of a drop in parasitism by *Apanteles* in September. In the Wayne–Wilson area the greatest production period was from mid-August to mid-September, with the peak about the first of September. The peak production weeks were similar to 1972, but production was greater due to a drop in parasitism by *W. manducae* in this area. Production in both areas again could have been reduced considerably by properly timed stalk destruction. However, overwintering production would have been relatively high in the Columbus–Bladen area as compared with other areas even with properly timed destruction.

An adequate cost-benefit analysis of the insect pest management program has not been made and would pose difficult problems. Nonmarket as well as monetary values, long- as well as short-term effects, and cost-benefit relationships to society as a whole as well as to individual growers are involved. The expenditure for scout salaries and travel was $4.70 per acre in 1973, which represented a reduction over 1972 but was higher than costs for scouting certain other crops. This high cost was related to the many small fields involved and the relatively large amount of data taken but could be reduced with modified scouting procedures.

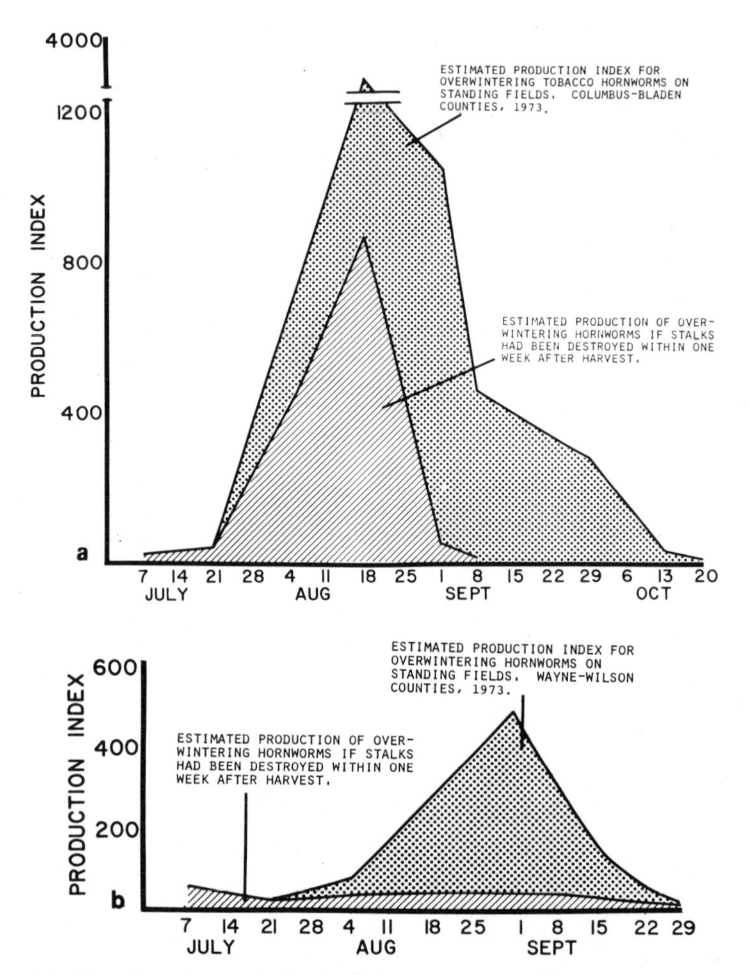

Figure 14. Estimated production index of overwintering tobacco hornworms in standing fields and estimated production if stalks had been destroyed within one week after harvest: (a) Columbus–Bladen and (b) Wayne–Wilson Counties, 1973.

Management for Control of Flue-Cured Tobacco Diseases

The pathologists also have devised an effective management program to combat tobacco diseases caused by an array of viruses, bacteria, fungi, and nematodes. These microscopic organisms contrast sharply with the more highly evolved insects, weeds, and vertebrate pests, but many biological and ecological principles apply to all.

Although the same fundamental phenomena and processes (such as reproduction, growth, development, movement, host relationships, mortality and survival, and genetic change) determine and regulate populations of all organisms, the microscopic size and astronomical numbers of pathogen propagules place important constraints on monitoring and studies of population dynamics. In addition, the phenomenal reproductive rates and extremely wide variation in dispersal mechanisms and capabilities among pathogens place practical temporal and spatial constraints on control options. As a consequence, it has been difficult to devise practical methods for measuring the population dynamics of viruses, bacteria, and fungi. Although population dynamics research is relatively new, pathologists are learning to solve some of the difficult problems. Techniques have been developed for a few fungal organisms, but progress has been notable in studies of nematodes. Thus, the nematologist and entomologist probably share more concepts regarding population management than either shares with other pathologists.

By the same token, the concept of managing a wide-area pest population is not readily applicable to all insect pests. If a crop in a particular field is to be protected from the ravages of a pest through population management, then management tactics must influence pest numbers and/or activities in that field during the period of protection. In other words, the components of the entire species population interacting with the field in question must be managed. Populations of pests, such as hornworms, that may move many miles from field to field cannot be managed by actions taken in one or a few fields by an individual grower. Thus, a few growers who fail to practice good postharvest sanitation may be responsible for a significant part of their neighbors' hornworm problems. On the other hand, the invasion pressure of many pathogens, with astronomical numbers of wind-borne spores or complex and poorly understood vector relationships, is much more complex in terms of source and timing. Population management as envisioned for hornworms has not been applied to relatively highly mobile plant pathogens up to this time. Populations of less mobile pests, such as wireworms and nematodes, where invasion rate is not as important in magnitude as with pests like hornworms, may be profitably studied in smaller spatial units and often can be managed in a practical manner by tactics applied to single fields. Thus, many *but not all* of the insect management programs require uniform grower compliance over wide geographic areas; whereas, most of the programs devised by pathologists to date are designed for the use of each grower independently.

As is the case with personnel of the tobacco insect pest management program, extension personnel in plant pathology have a valuable source of research information on which to base their tobacco disease control program. However, information from strictly research programs has been and continues to be augmented by an extensive on-the-farm program of testing various materials, tactics, and management systems. This program, which was conceptualized in cooperation with C. J. Nusbaum, was initiated and directed by F. A.

Todd. It is known to tobacco farmers in North Carolina as "Extension—Research on Wheels." No attempt will be made to present an adequate list of references to the information on which the action program in pathology is based. The few references cited may be no more than a useful entree to the literature. (See Lucas, 1965, for a general treatment.)

Of the many tobacco diseases, only four, each representing a different type of pathogen, will be included in this discussion:

Tobacco mosaic (Figure 15) is a widespread and costly disease causing mottling and burning of the leaves. It is caused by a virus and is transmitted mechanically. Horse nettle is the primary weed host known in North Carolina.

Bacterial (Granville) wilt is a serious threat in many but not all tobacco growing areas and causes the infected plant to wilt, typically beginning on only one side of the plant, but eventually causing plant death. It is caused by a soil-borne bacterium (*Pseudomonas solanacearum*). Plants in an advanced stage of infection with this bacterium bear a close resemblance to plants heavily infected with black shank (Figure 16).

Black shank (Figure 16) is caused by a soil-borne fungus (*Phytophthora parasitica* var. *nicotianae*) which apparently limits its attack to tobacco. The disease was first reported in North Carolina in 1931, and now it is considered unwise to use susceptible varieties in any tobacco growing area in the State.

Root knot (Figure 17) and other nematodes are present on practically all farms in the state. There are at least three groups of nematodes attacking tobacco in North Carolina, one of which is the root knot group, comprising five species, with *Meloidogyne incognito* considered the most important. Root knot nematodes occur at damaging levels in about 75% of the fields planted to tobacco, causing stunted plants, uneven stands (Figure 18) and loss in yield and quality.

The pathologists' action program has been named "R_x System Control—A Prescription for Flue-Cured Tobacco Disease" (Todd, 1971*b*). It recognizes the great variation in disease problems from farm to farm as well as the variation in what the growers can afford to do in controlling them. For example, one grower's total farm enterprise and land constraints may prohibit a rotation system of optimum value in disease control. Such a grower would then take counter measures, perhaps in selecting a different variety or in greater reliance on pesticides.

Six optional systems (Figure 19) have been devised for disease control, and the grower is advised to follow these five steps in selecting and implementing the system most suitable to his situation:

1. *Determine the diseases present.* The analysis is based primarily on the past incidence of disease in the fields involved as well as consultation with specialists.

Figure 15. Tobacco plant with tobacco mosaic. Note leaf mottling.

2. *Determine infestation level for each field.* Sampling problems make it impossible to base this evaluation on numbers of disease organisms per plant or area with respect to most pathogens, although such technology is available for certain nematodes. An experimental program of monitoring nematode populations as related to crop performance has been underway in North Carolina for several years (Barker and Nusbaum, 1971; Nusbaum and Ferris, 1973). The interactions between nematodes and other pathogens on tobacco increase the difficulties in predicting damage caused by a given pathogen (Powell, 1971; Kincaid *et al.*, 1970). At present, infestation levels are estimated on the basis of the percentage of plants infected during prior seasons. In rating infestation levels of Granville wilt or black shank, the following four categories are used:

Figure 16. Tobacco plants infected with the black shank fungus.

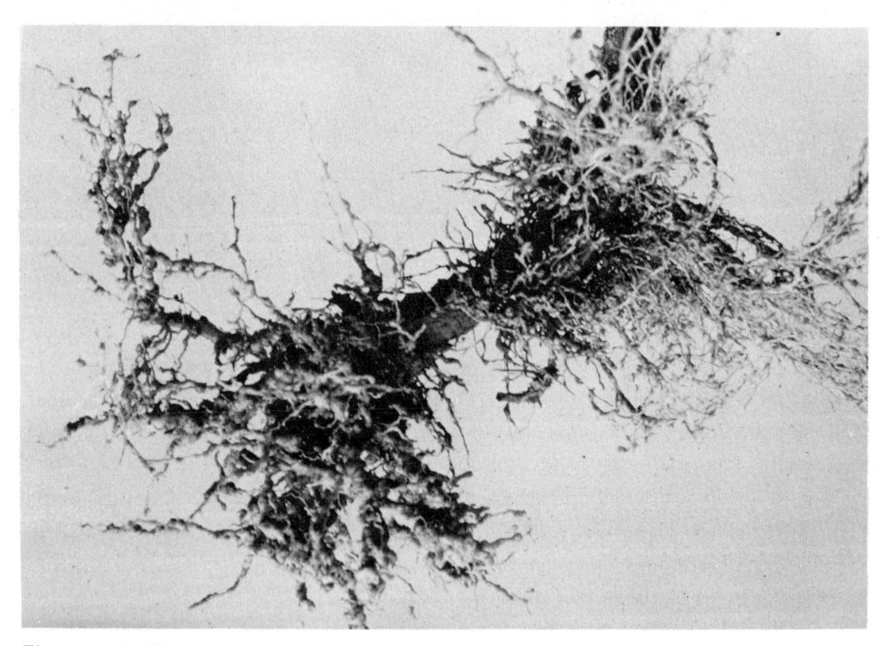

Figure 17. Root system of tobacco plant heavily infested with root knot nematodes. Note galls or swellings on roots.

Figure 18. Uneven stand of tobacco caused by root knot nematodes. Note stunted plants.

(1) very low—no disease noted, (2) low—less than 1% of plants died, (3) moderate—no more than 6% loss, (4) high—more than 6% loss. For root knot evaluation, root systems of plants after harvest and at the time of stalk destruction should be examined and categorized as follows on the basis of the percentage of root area with root knot galls: (1) very low—<10%, (2) low—10–15%, (3) moderate—25–50%, and (4) high—>50% (Figure 17). This index is but a rough guideline which can be misleading if not interpreted on the basis of phenology of the crop as well as related cultural practices such as stalk and root destruction.

3. *Consider production plans—is rotation practical?* This step requires a general analysis of the entire farm situation in order to decide optimal spatial distribution of crops and rotation systems.

4. *Selecting a system.* Each of the six systems suggested is for a different situation, so that the grower can select the one which most closely meets his requirements.

5. *Learn details of practices included in the system selected.*

The following practices are integrated to various degrees in the management of "prescription" systems:

RELATIVE INFESTATION LEVELS OF THREE MAJOR TOBACCO DISEASES

A	B	C	D	E	F
BLACK SHANK, WILT-HIGH	BLACK SHANK, WILT-LOW TO MODERATE	BLACK SHANK, WILT-LOW TO MODERATE	BLACK SHANK, WILT-LOW	BLACK SHANK, WILT-VERY LOW	BLACK SHANK, WILT-VERY LOW
ROOT-KNOT-LOW TO HIGH	ROOT-KNOT-LOW TO HIGH	ROOT-KNOT-LOW TO HIGH	ROOT-KNOT LOW TO HIGH	ROOT-KNOT-LOW TO HIGH	ROOT-KNOT-VERY LOW

SUGGESTED MANAGEMENT SYSTEMS FOR EACH INFESTATION LEVEL

A	B	C	D	E	F
CROPPING SYSTEM-2 OR 3 YEAR ROTATION	CROPPING SYSTEM-NO ROTATION	CROPPING SYSTEM-2 YEAR ROTATION	CROPPING SYSTEM-2 YEAR ROTATION	CROPPING SYSTEM-2 YEAR ROTATION	CROPPING SYSTEM-2 OR 3 YEAR ROTATION
VARIETY-HIGH BLACK SHANK, WILT, ROOT-KNOT	VARIETY-HIGH BIACK SHANK, WILT, ROOT-KNOT	VARIETY-HIGH BLACK SHANK	VARIETY-LOW TO MODERATE	VARIETY-LOW TO MODERATE	VARIETY-LOW TO MODERATE
MULTI PURPOSE CHEMICAL	MULTI PURPOSE CHEMICAL	NEMATICIDE	MULTI PURPOSE CHEMICAL	NEMATICIDE	NO CHEMICAL

NOTE: FOUR PRACTICES INCLUDING OPERATION R-6-P, PLANT BED DISEASE CONTROL, MOSAIC PREVENTION AND BROWN SPOT CONTROL ARE IMPORTANT AND SHOULD BE INCLUDED IN EACH SYSTEM.

Figure 19. Summary of systems suggested for growers to use in combating tobacco disease problems of differing intensities.

1. *Stalk destruction and plowing out tobacco stubble*. This is aimed primarily at nematodes, mosaic, and brown spot. This practice is called R6P (reduced six pests) because early stalk destruction also reduces overwintering hornworms, budworms, and fleabeetles.

2. *Rotation with resistant crops*. This is aimed particularly at black shank, nematodes, Granville wilt, and virus. A selection of a fairly large array of resistant crops is available according to locality and disease complex present.

3. *Selecting varieties resistant to the disease complex present*. Varieties are available with various levels of resistance to root knot, black shank, and Granville wilt; however, varieties must also be selected with regard to other factors varying from farm to farm.

4. *Control plant bed diseases*. This practice involves the use of fungicides for prevention of blue mold, anthracnose, and damping off. The control of these diseases is essential for production of an abundant supply of healthy transplants.

5. *Prevent spread of mosaic*. Mosaic can be spread easily by handling plants at transplant time if the virus is present. In addition, the virus is present in much of the manufactured tobacco products. Thus, smoking and chewing should be avoided in plant beds and tobacco fields, and periodic rinsing of hands in milk while handling transplants aids in preventing spread (Hare and Lucas, 1959).

6. *Select a chemical treatment to fit the disease problem of the field as well as the other practices selected.* Two groups of chemicals are available: nematicides and multipurpose chemicals. The latter are effective not only against nematodes but also reduce black shank, wilt, and certain other soil-borne pests.

Stalk destruction, the production of healthy transplants and the prevention of mosaic spread are practices common to all of the six systems suggested (Figure 19). Cropping systems, selection of varieties, and chemical soil treatments, however, vary among the systems, and decisions on these latter practices should be made on the basis of infestation levels. For example, the situation depicted in the far left column of Figure 19 involves a field where levels of all three diseases are high, and the practices recommended are two- or three-year rotation, a variety with high resistance to all three diseases listed, and a soil application of a multipurpose chemical. In the second and third situation noted in Figure 19, black shank and wilt are low to moderate and nematodes are low to high. In such a situation, a farmer with limited land for rotation might choose system B with no rotation or if he has land available for rotation, he might choose system C with a two-year rotation, a greater choice of varieties and a soil application of a nematicide rather than the more expensive and wider spectrum multipurpose treatment. At the other end of the infestation spectrum (last column in Figure 19), all diseases are low, and the farmer might choose system F, with a two- or three-year rotation, full freedom in variety selection, and no chemical treatment.

Where all diseases are heavy, no one practice will give satisfactory results and profitable production can be realized only if several practices are integrated. A seven-year study, conducted in a field heavily infested with all three major diseases, yielded information (Table V) relative to the contribution of

TABLE V. Value of System A [a] Practices under Heavy Disease Pressure (Seven-Year Study)

Practices	Increase, $	Total value, $
None (check plots)	—	424
Stalk destruction	288	
Rotation	447	
Resistant variety	380	
Soil chemicals	558	
All practices integrated		2097

[a] See Figure 19 for description of components of system A.

Figure 20. Tobacco plots in which system A practices were used (a) and in which none of these practices were followed (b).

each practice to the value of the crop (Todd, 1971*c*). The appearance of tobacco in plots where system A was used as compared with a check plot, where none of the practices were followed, is illustrated in Figures 20a and 20b, respectively.

Integration of Insect and Disease Management

The two action programs, as described above, now fit into tobacco production with little noticeable negative interaction among the tactics involved. Although the entomologist and pathologist share many basic principles useful in control, the relative importance of specific strategies and tactics varies appreciably according to the type of pest involved.

As shown in Table VI, resistant varieties and rotation are important tactics for controlling diseases and nematodes but not insects (currently). Some variation in insect response to various commercial varieties (many developed for disease control) has been noted (Thurston, 1961; Elsey and Rabb, 1967, Girardeau, 1968, 1969); however, none of the varieties have induced a recognized insect problem. Entomologists may wish to have more input into variety selection in the future, and the potential of rotation as a tactic for wireworm control has not been explored adequately.

The greatest use of "insurance" treatments with pesticides is for wireworms and nematodes. Currently, there are relatively few pesticidal applications on field tobacco for disease control other than those caused by nematodes,

TABLE VI. Relative Importance of Strategies and Tactics for Controlling Insect Pests, Nematodes, and Other Diseases in Fields of Tobacco.

	Relative significance			
Strategies and tactics	Foliage-feeding insects	Soil insects	Nematodes	Other diseases
Eliminate susceptibility				
Resistant varieties	None	None	High	High
Suppress pest populations				
Insurance treatments—pesticides	Low	Moderate	Moderate	Low
Selective treatments—pesticides				
Based on previous history	Low	Moderate	High	Low
Based on current infestation	High	Low	Low [a]	Low
Rotation	Low	None	High	High
Destruction of crop residues	High	None	High	High

[a] Nematode assays are being used to assist tobacco growers in determining control measures needed. This program is being expanded presently by the North Carolina Department of Agriculture.

and the trend in the use of foliar insecticides is toward a decrease in the insurance treatment approach.

Selective treatments with chemicals for disease control are primarily based on previous history; whereas, with foliar insect pests, decisions to treat can be made for most species on the basis of field counts coupled with the use of economic thresholds. (Economic thresholds in use are relatively crude and set so low that a large margin for error exists.) Monitoring disease organisms (other than nematodes) generally is much more difficult and complex than monitoring insects; thus, decisions to use preplant chemicals against disease organisms are made chiefly on the basis of "symptom monitoring" the previous crop in a specific field. Negative interactions between the use of insecticides, fungicides, and nematicides have not been noticed except in cases where chemicals have produced phytotoxic effects, and these cases have been minimized by careful screening. However, the advent of a number of multipurpose pesticides and pesticidal mixtures for use against nematodes, other soil-borne pathogens, and insects may pose problems in integrating the inputs of entomologists and pathologists. If such materials are recommended and used on large acreages in a routine manner without regard to need (i.e., counter to the economic threshold principle), problems of resistance and/or disruption may occur.

The last tactic listed in Table VI, destruction of crop residues, is of importance in controlling insect, nematode, and certain other soil-borne pests. Although there are no negative interactions involved between recommended postharvest practices as viewed by entomologists and pathologists, requirements for optimum results are somewhat different. The entomologist seeks to eliminate foliage and flowers as soon as possible after harvest. This is sometimes accomplished with a heavy treatment of MH. However, the pathologist is concerned about reproduction of nematodes and soil-borne pathogens in the roots of the plant. Therefore, merely cutting stalks is not enough; the roots must be plowed out and destroyed.

Seeking a Practical Level of Sophistication

I should like to close with a philosophical observation on seeking a practical level of sophistication in modeling with respect to tobacco pest management.

The ability of a good model may be expressed in terms of more effective planning in research, greater understanding of cause and effect relationships, reduced counterintuition, and more effective management of the systems modeled in terms of net gain. A good model must have appropriate levels of generality, reality, and predictability. Our current tobacco pest management program rates rather well in terms of these criteria. It has generality in that it can be applied to farms throughout North Carolina; it has reality in that it can be used successfully for the solution of problems in the real world; it has reasonable predictiveness in that the recommended systems provide a high level of reliability.

A diagrammatic representation (Figure 21) of the relationship between accuracy and predictiveness of our conceptual models of tobacco pest management demonstrates in a very generalized way our ability to predict correct decisions. Keep in mind these decisions are often multiple choice and not simply yes or no. The dotted horizontal line represents an estimated, minimal, acceptable accuracy for our pest management decision-making process. The curved line indicates that with our present conceptual models and data acquisition systems we can predict with acceptable accuracy on a short-term basis, but with less acceptable accuracy on a long-term basis. For individual fields, the need for foliar applications of insecticides can be predicted one to several weeks in advance, and the need for rotation or soil treatment for certain diseases as much as seven to eight months in advance.

More sophisticated mathematical modeling theoretically could improve the accuracy and predictiveness of our program both spatially and temporally. Such improvement would require a more realistic model, with account taken of more factors and factor interaction than we presently recognize and more input of information. If we were to obtain greater accuracy and the ability to predict a given level of accuracy for a longer period, the curve in this figure would be raised with some modification in shape. However, this would be costly

An examination of costs and benefits of this more advanced modeling, as depicted in Figure 22, shows that costs and benefits increase with accuracy, but

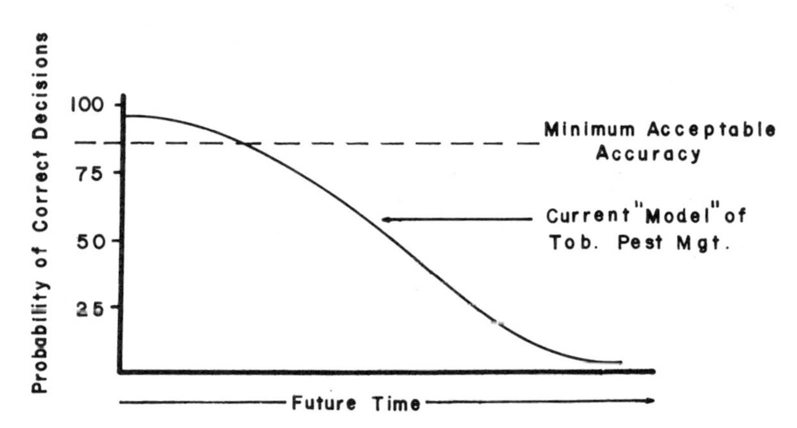

Figure 21. Generalized representation of accuracy and predictiveness of the current conceptual model of tobacco pest management.

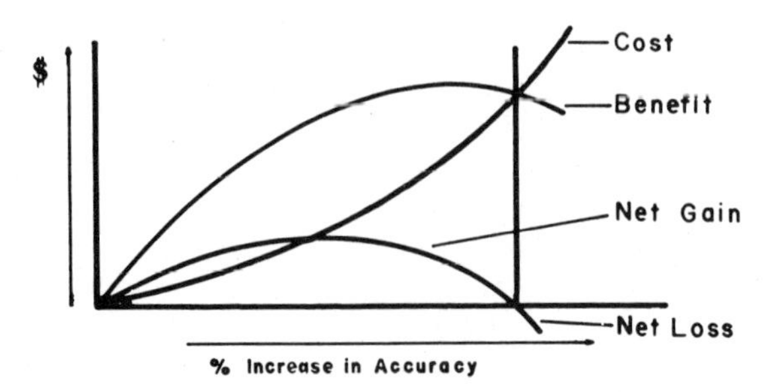

Figure 22. Cost–benefit relationship in increasing model sophistication and accuracy.

that benefits reach a point of diminishing returns. At the point of intersect of the two lines, net gain = 0, and if modeling continues after this point, it represents a net loss. Obviously Figure 22 is merely a generalization of the relationships involved, but it would be difficult to insert meaningful figures into such a cost-benefit formula. There is no serious question as to the wisdom of applying the systems approach more effectively in all the disciplines and specialty areas involved in tobacco production. Such an approach highlights some obvious research needs, such as developing more satisfactory monitoring techniques, more accurate and practical economic thresholds, in-depth studies of the population dynamics of key pests, and a more useable understanding of the indirect as well as direct effects of actual and potential major changes in tobacco production technology. In the latter category, no-tillage and the widespread use of soil pesticides are two examples of changes that may well alter the insect–disease situation significantly. For optimum development and maintenance of tobacco production, the modeling approach must be used along with in-depth, high-quality research by capable specialists of traditional disciplines. It is not a question of one approach versus the other; both are essential. However, we must keep in mind that modeling is merely a tool, and if misapplied can result in wasted resources and could serve to discredit modeling as one of our most useful tools in solving complex problems. A rather simple modeling approach may suffice for many problems, and using highly sophisticated models in these instances would be inappropriate. Given the present level of organization in tobacco production, it might be difficult to justify highly sophisticated models. But studies of the subsystems of tobacco production, such as the life systems of certain pests, will require advanced techniques of modeling.

As pest management specialists and agriculturists, our major objective is to devise and maintain practical pest management systems rather than to construct complex models which may or may not be justified economically. Where they are justified, however, we should use them in the context of our major objective. On the other hand, we also have a responsibility to support the basic research that undergirds pest management and agriculture, and the research worker is now finding modeling to be an essential tool with respect to many basic problems. Thus, modeling undoubtedly will play an important although changing role in our efforts to improve tobacco production and minimize deleterious environmental effects.

Literature Cited

Allen, N., Hodge, C. R., and Early, J. D., 1952, Insecticides may cause an objectionable flavor or odor to tobacco, 64th Annual Report, South Carolina Experiment Station, p. 84.

Allen, N., Kinard, W. S., and Creighton, C. S., 1961, *Bacillus thuringiensis* controls the tobacco budworm, *Tob. Sci.* **5**:58–62.

Barker, K. R., and Nusbaum, C. J., 1971, Diagnostic and Advisory Programs, *in: Plant Parasitic Nematodes* (B. M. Zuckerman, W. F. Mai, and R. A. Rohde, eds.), Academic Press, New York, Vol. I, pp. 281–301.

Bowery, T. G., Evans, W. R., Guthrie, R. E., and Rabb, R. L., 1959, Insecticide residues on tobacco, *J. Agr. Food Chem.* **7(10)**:693–702.

Bowery, T. G., Gatterdam, P. E., Guthrie, F. E., and Rabb, R. L., 1965, Metabolism of insecticide residues—Fate of inhaled C^{14}-TDE in rabbits, *J. Agr. Food Chem.* **13**:356–359.

Bradley, R. H. E., and Ganong, R. Y., 1957, Three more viruses borne at the stylet tips of the aphid *Myzus persicae* (Suz.), *Can. J. Micro.* **3**:669–70.

Clayton, E. E., Graham, T. W., Todd, F. A., Gaines, J. G., and Clark, F. A., 1958, Resistance to the root knot disease of tobacco, *Tob. Sci.* **2**:53–63.

Creighton, C. S., Kinard, W. S., and Allen, N., 1961, Effectiveness of *Bacillus thuringiensis* and several chemical insecticides for control of budworms and hornworm on tobacco, *J. Econ. Entomol.* **54**:1112–1114.

Davide, R. G., and Triantaphyllou, A. C., 1968, Influence of the environment and sex differentiation of root-knot nematodes. III. Effect of foliar application of maleic hydrazide, *Nematologica* **14**:37–46.

DeLoach, C. J., and Rabb, R. L., 1971, Life History of *Winthemia manducae* (Diptera: Tachinidae), a parasite of the tobacco hornworm, *Ann. Entomol. Soc. Am.* **64(2)**:399–409.

DeLoach, C. J., and Rabb, R. L., 1972, Seasonal abundance and natural mortality of *Winthemia manducae* (Diptera: Tachinidae) and degree of parasitization of its host, the tobacco hornworm, *Ann. Entomol. Soc. Am.* **65(4)**:779–790.

Domanski, J. J., Laws, J. M., Haire, P. L., and Sheets, T. J., 1973, Insecticide residues on U.S. tobacco products, *Tob. Sci.* **17**:80–81.

Dominick, C. B., 1968, Evaluation of experimental insecticides for control of hornworms on tobacco, *J. Econ. Entomol.* **61**:483–484.

Ellis, H. C., and Ganyard, M. C., Jr., 1973, North Carolina tobacco pest management: Third annual report, Unnumbered publication of North Carolina Agricultural Extension Service. 37 pp.

Ellis, H. C., Ganyard, M. C., Jr., Singletary, H. M., and Robertson, R. L., 1972, North Carolina tobacco pest management: Second annual report, Unnumbered publication of North Carolina Agricultural Extension Service. 67 pp.

Elsey, K. D., 1972, Predation of eggs of *Heliothis* spp. on tobacco, *Environ. Entomol.* **1(4)**:433–438.

Elsey, K. D., 1973*a*, *Jalysus spinosus:* Effect of insecticide treatment on this predator of tobacco pests, *Environ. Entomol.* **2(2):**240–243.

Elsey, K. D., 1973*b*, *Jalysus spinosus:* Spring biology and factors that influence occurrence of the predator on tobacco in North Carolina, *Environ. Entomol.* **2(3):**421–425.

Elsey, K. D., and Rabb, R. L., 1967, Biology of the cabbage looper on tobacco in North Carolina, *J. Econ. Entomol.* **60:**1636–1639.

Elsey, K. D., and Stinner, R. E., 1971, Biology of *Jalysus spinosus,* an insect predator found on tobacco, *Ann. Entomol. Soc. Am.* **64(4):**779–783.

Fulton, B. B., 1940, The hornworm parasite, *Apanteles congregatus* (Say) and the hyperparasite, *Hypopteromalus tabacum* (Fitch), *Ann. Entomol. Soc. Am.* **33(2):**231–44.

Ganyard, M. C., Jr., Ellis, H. C., and Singletary, H. M., 1971, North Carolina tobacco pest management: First annual report, Unnumbered publication of North Carolina Agricultural Extension Service. 36 pp.

Girardeau, J. H., 1968, The influence of variety in flue-cured tobacco on damage by the tobacco budworm, *J. Ga. Entomol. Soc.* **3:**51–55.

Girardeau, J. H., 1969, The effect of level of fertilization and variety of flue-cured tobacco on damage by the tobacco budworm, *J. Ga. Entomol. Soc.* **4:**85–89.

Guthrie, F. E., 1968, The nature and significance of pesticide residues on tobacco and in tobacco smoke, *Beitrage Tabakforschung* **4:**22.

Guthrie, F. E., 1973, Pending legislative restrictions on the use of agricultural chemicals on tobacco, *Beitrage Tabakforschung* **7:**195–202.

Guthrie, F. E., McCants, C. B., and Small, H. G., Jr., 1959*a,* Arsenic content of commercial tobacco, 1917–1958, *Tob. Sci.* **3:**62–64.

Guthrie, F. E., Rabb, R. L., and Bowery, R. G., 1959*b,* Evaluation of candidate insecticides and insect pathogens for tobacco hornworm control 1956–1958, *J. Econ. Entomol.* **52:**798–804.

Guthrie, F. E., Rabb, R. L., Bowery, T. G., Lawson, F. R., and Baron, R. L., 1959*c,* Control of hornworms and budworms on tobacco with reduced insecticide dosage, *Tobacco* **148(2):**20–23.

Guthrie, F. E., Rabb, R. L., and Mount, D. A., 1963, Distribution and control of cyclodiene-resistant wireworms attacking tobacco in North Carolina, *J. Econ. Entomol.* **56:**7–10.

Hare, W. W., and Lucas, G. B., 1959, Control of contact transmission of tobacco mosaic virus with milk, *Plant Dis. Reptr.* **43:**152–154.

Kinard, W. S., Henneberry, T. J., and Allen, N., 1972, Tobacco stalk cutting: Effect on insect populations, *J. Econ. Entomol.* **65:**1417–1421.

Kincaid, R. R., Martin, F. G., Gammon, N., Jr., Breland, H. L., and Pritchett, W. L., 1970, Multiple regression of tobacco black shank, root knot, and coarse root indexes on soil pH, potassium, calcium, and magnesium, *Phytopathology* **60:**1513–1516.

Lawson, F. R., 1959, The natural enemies of the hornworms on tobacco (Lepidoptera: Sphingidae), *Ann. Entomol. Soc. Am.* **52(6):**741–55.

Lawson, F. R., Corley, C., and Schechter, M. S., 1964, Insecticide residues on tobacco during 1962, *Tob. Sci.* **8:**110–112.

Lawson, F. R., and Rabb, R. L., 1964, Factors controlling hornworm damage to tobacco and methods of predicting outbreaks, *Tob. Sci.* **8:**145–149.

Lawson, F. R., Rabb, R. L., Guthrie, F. E., and Bowery, T. G., 1961, Studies of an integrated control system for hornworms in tobacco, *J. Econ. Entomol.* **54:**93–97.

Lucas, G. B., 1965, *Diseases of Tobacco* (2nd ed.) The Scarecrow Press, Inc., New York. 778 pp.

Madden, A. H., and Chamberlin, F. S., 1945, Biology of the tobacco hornworm in the southern cigar-tobacco district, *USDA Tech. Bull.* **896.** 51 pp.

McNeil, J. N., and Rabb, R. L., 1973*a,* Life histories and seasonal biology of four hyperparasites of the tobacco hornworm, *Manduca sexta* (Lepidoptera: Sphingidae), *Can. Entomol.* **105(8):**1041–1052.

McNeil, J. N., and Rabb, R. L., 1973*b*, Physical and physiological factors in diapause initiation of two hyperparasites of the tobacco hornworm *Manduca sexta*, *J. Insect Physiol.* **19:**2107–2118.

Metcalf, Z. P., 1909, Insect enemies of tobacco, *N.C. Dept. Agr. Spec. Bull.* **70.** 72 pp.

Mistric, W. J., Jr., and Pittard, W. W., 1973, Damage to flue-cured tobacco by tobacco budworm and corn earworm along and combined at various infestation densities, *J. Econ. Entomol.* **66:**232–235.

Mistric, W. J., Jr., and Smith, F. D., 1969, Behavior of tobacco budworm larvae on flue-cured tobacco and possibilities of improving the effectiveness of insecticidal treatments applied mechanically for control. *J. Econ. Entomol.* **62:**16–21.

Mistric, W. J., Jr., and Smith, F. D., 1973*a*, Tobacco hornworm: methomyl, monocrotophos, and other insecticides for control on flue-cured tobacco, *J. Econ. Entomol.* **66:**581–583.

Mistric, W. J., Jr., and Smith, F. D., 1973*b*, Tobacco budworm: Control on flue-cured tobacco with certain microbial pesticides, *J. Econ. Entomol.* **66:**979–982.

Moore, E. L., Powell, N. T., Jones, G. L., and Gwynn, G. R., 1962, Flue-cured tobacco variety NC-95, resistant to root-knot, black-shank, and the wilt disease, *N.C. Agr. Exp. Sta. Bull.* **419.** 18 pp.

Neunzig. H. H., 1969, The biology of the tobacco budworm and the corn earworm in North Carolina with particular reference to tobacco as a host, *N.C. Agr. Exp. Sta. Tech. Bull.* **196.** 76 p.

North Carolina State College, Entomology Faculty, 1958, Insecticide residues as a source of off flavor in tobacco, *Tob. Sci.* **2:**90–94.

Nusbaum, C. J., 1958, The response of root-knot-infected tobacco plants to foliar applications of maleic hydrazide, *Phytopathology* **48:**344 (Abstr).

Nusbaum, C. J., and Ferris, H., 1973, The role of cropping systems in nematode population management, *Ann. Rev. Phytopath.* **11:**423–440.

Nusbaum, C. J., and Todd, F. A., 1970, The role of chemical soil treatments in the control of nematode-disease complexes of tobacco, *Phytopathology* **60:**7–26.

Nusbaum, C. J., Guthrie, F. E., and Rabb, R. L., 1961, The incidence of stem roots in tobacco transplants in relation to wireworm injury, *Plant Dis. Reptr.* **45(3):**225–226.

Powell, N. T., 1971, Interactions of plant parasitic nematodes, *in: Plant Parasitic Nematodes* (B. M. Zuckerman, W. F. Mai, and R. A. Rhode, eds.) Academic Press, New York, Vol. II, pp. 119–316.

Rabb, R. L., 1960, Biological studies of *Polistes* in North Carolina (Hymenoptera: Vespidae), *Ann. Entomol. Soc. Am.* **53(1):**111–121.

Rabb, R. L., 1966, Diapause in *Protoparce sexta* (Lepidoptera: Sphingidae), *Ann. Entomol. Soc. Am.* **59(1):**160–165.

Rabb, R. L., 1969, Environmental manipulations as influencing populations of tobacco hornworms, Proceedings of the Tall Timbers Conference on Ecological Animal Control by Habitat Management, No. 1.

Rabb, R. L., 1971, Naturally-occurring biological control in the eastern United States, with particular reference to tobacco insects, *in: Biological Control* (C. B. Huffaker, ed.). Plenum Press, New York.

Rabb, R. L., and Bradley, J. R., Jr. 1968, The influence of host plants on parasitism of eggs of the tobacco hornworm, *J. Econ. Entomol.* **61:**1249–1252.

Rabb, R. L., and Guthrie, F. E., 1956, Tobacco hornworm control experiments in North Carolina, 1953–1955, *J. Econ. Entomol.* **49:**818–820.

Rabb, R. L., and Guthrie, F. E., 1964, Resistance of tobacco hornworms to certain insecticides in North Carolina, *J. Econ. Entomol.* **57:**995–996.

Rabb, R. L., and Lawson, F. R., 1957, Some factors influencing the predation of *Polistes* wasps on the tobacco hornworm, *J. Econ. Entomol.* **50:**778–784.

Rabb, R. L., and Thurston, R., 1969, Diapause in *Apanteles congregatus, Ann. Entomol. Soc. Am.* **62:**125–128.

Rabb, R. L., Steinhaus, E. A., and Guthrie. F. E., 1957, Preliminary tests using *Bacillus thuringiensis* Berliner against hornworms, *J. Econ. Entomol.* **50:**259–262.

Rabb, R. L., Neunzig, H. H., and Marshall, H. V., Jr., 1964, Effect of certain cultural practices on the abundance of tobacco hornworms, tobacco budworms, and corn earworms on tobacco after harvest, *J. Econ. Entomol.* **57:**791–792.

Reagan, T. E., Rabb, R. L., and Collins, W. K., 1974, Tobacco budworm: Influence of early topping and sucker control practices on infestations in flue-cured tobacco, *J. Econ. Entomol.* **67(4):**551–552.

Seltmann, Heinz, 1971, Modern methods of tobacco sucker control, *in:* Proceedings Actes Du Congress Kongressbericht, Hamburg, 1970, pp. 77–84.

Sheets, T. J., Smith, J. W., and Jackson, M. D., 1968, Pesticide residues in cigarettes, *Tob. Sci.* **12:**66.

Tappan, W. B., Wheeler, W. B., and Lundy, H. W., 1973, Methomyl residues on cigar-wrapper and flue-cured tobaccos in Florida, *J. Econ. Entomol.* **66:**197–198.

Thurston, R., 1961, Resistance in *Nicotiana* to the green peach aphid and some other tobacco insect pests, *J. Econ. Entomol.* **54:**940:–949.

Todd, F. A., 1971*a,* Model systems for integrated use of chemicals, resistant varieties and cultural practices for tobacco disease control, *in:* Proceedings Actes Du Congress Kongressbericht, Hamburg, 1970, pp. 90–104.

Todd, F. A., 1971*b,* System control: A prescription for flue-cured tobacco disease, *N.C. Agr. Ext. Serv. Cir.* **530.** 15 pp.

Todd, F. A., 1971*c,* Extension-research on wheels, flue-cured tobacco summary report of 1971 data, Plant Pathology Information Note No. 179, pp. 123–127.

Von Ramm, C., Lucas, G. B., and Marshall, H. V., Jr., 1962, The effect of maleic hydrazide on brown spot incidence of flue-cured tobacco, *Tob. Sci.* **6:**100–103.

J. H. Webber, 1956, Arsenic in cigarette tobacco, *J. Sci. Food Agr.* **8:**490–491.

VIII

Systems Approach to Cotton Insect Pest Management

D. W. DeMichele and Dale G. Bottrell

Cotton is a major world agricultural crop cultivated for the harvest of lint fibers utilized extensively in the manufacture of apparel and household and industrial goods. Although linked biologically to the production of cotton fibers, cottonseeds are more than a mere byproduct of the cotton harvest. Seeds ginned from the lint fibers are processed commercially for use in animal feeds, foods for human consumption, and concoctions used in the preparation of these foods as well as numerous other home and industrial products.

Cotton is an important mainstay to the economy of many countries and contributes significantly to a productive and balanced world agriculture. The crop is particularly important in shaping the future of many developing countries whose export earnings are largely derived from the sale of harvested cotton (Reynolds *et al.*, 1975). The importance of cotton to the future of a developed country, such as the United States, should be apparent if we examine a few facts illustrating the present economic and social values of the crop.

Cotton is the fifth most valuable crop in the United States (Ranney, 1973). It accounts for more than 50% of the income from agricultural crop production in more than half of the 19 states in which it is grown (Reynolds *et al.*, 1975). This amounts to more than $2.25 billion in farm income to some 200,000 growers (Ranney, 1973). More than $13 billion is invested in 13 million acres of land and in the equipment to grow cotton, with billions more invested in gins, oil mills, warehouses, textile plants, and merchandising establishments. The incomes of over 5.2 million Americans derive wholly or in part from cot-

D. W. DEMICHELE · Department of Industrial Engineering, Texas A & M University. DALE G. BOTTRELL · Department of Entomology, Texas A & M University, College Station, Texas. Presently, private entomological consultant, 542 Piezzi Road, Santa Rosa, California 95401.

ton, and 12 million are employed by industries related to cotton. Additionally, as a high-valued export item ($748 million from lint alone in fiscal 1972–1973), cotton contributes significantly to the nation's trade balance each year (Ranney, 1973).

Obviously, there are many factors that will influence the future contribution of cotton to a continued productivity and balance of economies in developed countries and in shaping the development of emerging countries. Changes in government actions on farm programs, environmental problems, worker safety, and health and consumer protection, as well as changes in technology and the competitive role of synthetic fibers will undoubtedly have an impact on this future. While the future role of other agricultural crops will also be influenced by some of these changes, cotton is quite unique in one respect. The harvested cotton fiber is an industrial raw material that is challenged by competition from products of science. The present high costs and shortages of petroleum from which these products are derived has provided some relief from this competition since the production of cotton fiber requires only about one-fifth as much fossil fuel energy as manmade fiber substitutes (Ranney, 1973). On the other hand, research expenditures by the synthetic fiber industry are much greater than expenditures for cotton research: $150 million per year compared to $40 million per year in the United States (Ranney, 1973).

Systems Approach to Increased Cotton Efficiency

It is apparent that increased efficiencies are needed in all phases of cotton production, handling, processing, and marketing if the crop is to survive competition from manmade products and the many other countermanding factors. While improved efficiency will help meet competition from manmade products, increased efficiencies will be no less important in other major areas: improved fiber quality; development of new and improved fiber products; enhancement of cottonseed quality; and solutions to problems of worker safety and health, consumer protection and environmental quality (Ranney, 1973). That is, we need to examine critically the whole system surrounding cotton from the time it is planted until the harvested products have reached the consumer. Then, appropriate adjustments need to be made to realign each of the system's interlocking components or subsystems to maximize the efficiency of the system as a whole. Clearly, this approach must use a research strategy that anticipates integration of the results at the end of the research program. Otherwise, as Watt (1966) pointed out, we may complete the research only to find that the fragmented results obtained from various components cannot be fitted together as a meaningful "whole." This horrendous outcome may be avoided by designing the research program in terms of a conceptual model comprising submodels of the various parts (Watt, 1966). The philosophy of this approach is referred to as "systems science," and its problem solving methodology is commonly referred to as "the systems approach" (Jenkins, 1969).

Systems science has emerged from attempts to study and control larger and more complex systems (Witz, 1973). While the systems approach has only recently been utilized extensively in agricultural research, it has been used for some time in the fields of naval logistics, missle constructions, space exploration, etc. Its recent entry into agriculture has come about as the result of the increasing emphasis on interdisciplinary research in most agricultural fields and the agriculturist's greater acceptance of computer technology.

It is beyond the scope of this chapter to review the gamut of current work in cotton which has utilized the systems approach. For an overview of the current trends of research in the whole field of cotton research in the United States, we refer the reader to The 1973 Cotton Research Task Force Report (Ranney, 1973) and recent volumes of the Proceedings of the Beltwide Cotton Production Research Conferences published by the National Cotton Council. It is obvious from the discussions within these references that the principal public agricultural research agencies of this nation (the Agricultural Research Service–United States Department of Agriculture (ARS–USDA) and the Land-Grant University Agricultural Experiment Stations) have placed a high priority on adopting systems approaches in cotton research through the efforts of interdisciplinary teams of biological and physical scientists, economists, and others. Benefits have already emerged from this interdisciplinary systems approach. A major one is the improved network for communication among the various disciplines engaged in cotton research, which in turn has catalyzed greater understanding among disciplines and the team approach toward mission-oriented objectives.

There presently are several large-scale interdisciplinary research projects in the United States and other countries that have adopted the systems approach to the management of cotton insect pests. Some of these projects have shown good progress and undoubtedly have shaped an approach to cotton pest management that shall persist for many years to come. We will not attempt to discuss all of these projects or any one in great detail. Rather, we will describe the basic problems in cotton supporting the rationale for adopting a systems approach to pest management, cite the problems and progress in computer modeling the various components of the cotton ecosystem, and discuss the use of models and systems analysis in making pest control decisions.

Most of our discussion centers on the current work of the project, "The Principles, Strategies, and Tactics of Pest Population Regulation and Control in the Cotton Ecosystem" which is a contributing component of the much larger multicrop project commonly called the "Huffaker" or US–IPM Project. A major objective of this project is to develop in each of the areas represented* an ecologically based system of cotton insect pest management that will optimize costs-benefits over the long-term to both the farmer and to society.

The Argument for the Systems Approach

Several recent articles have described the general problems of cotton insect control in the United States and have presented the underlying justification and rationale behind the recent trend toward ecologically based, integrated pest management systems as opposed to the unilateral, insecticide-based programs of the past (Adkisson, 1973*a, b;* Smith and Reynolds, 1972; Smith and Falcon, 1973; Smith and Huffaker, 1973; Smith *et al.,* 1974; Reynolds *et al.,* 1975; Huffaker and Croft, 1975; Huffaker and Smith, 1973; and van den Bosch *et al.,* 1971). All of these articles point out the shortcomings of relying solely on insecticides for control of cotton insect pests and cite examples of some of the serious problems which have resulted from the misuse of insecticides.

Presently, there are 24 species of insect and spider mite pests of cotton which have developed resistance to one or more insecticides (Twenty-seventh Conference, 1974). Several other pest species are strongly suspected of having developed resistance. Resistance is widespread among most of the major insect pests as well as within populations of some of the potential and occasional insect pests. In most cases, these insects are resistant only to the organochlorines, but in other cases organophosphorus (OP) resistance has developed where these materials have been used extensively as substitutes against the organochlorine-resistant pests. The OP pesticides are less persistant and, thus, have to be applied more frequently than the organochlorines with often resulting higher costs to the user, increased problems of pesticidal drift, more frequent ecological upsets, and increased problems of human poisoning. The long-term result is often disaster as exemplified by the case described by Adkisson (1973*b*) for south Texas and northeast Mexico following the development of resistance in the tobacco budworm, *Heliothis virescens* (F). This pest, first resistant to the organochlorines and later resistant to the substitute carbamates and OP's, has made cotton production unprofitable in certain parts of Mexico and threatens to

* The cotton project is underway at the land-grant universities of Arkansas, California, Mississippi, and Texas with collaborating ARS–USDA (Agricultural Research Service–United States Department of Agriculture) laboratories at several locations. Refer to Reynolds (1973), Bottrell (1973), Phillips (1973), and Harris (1973) for a general description of work underway at the individual states; Adkisson (1973*a, b*), Huffaker and Smith (1973), Smith and Huffaker (1973), Smith *et al.* (1974), and Huffaker and Croft (1975) for discussions pertaining to the overall national project.

cause economic disaster in all of the south Texas–northeast Mexico area where cotton historically has been an important economic crop.

The pesticidal-induced problems mentioned above, the economic disaster situations that often have accompanied these problems, and the increased government constraints on grower use of insecticides have been the predominant reasons behind the recent initiative toward development of more ecologically acceptable methods of cotton insect control. There are increased research efforts toward developing new and improved tactics for controlling major pests such as the boll weevil, *Anthonomus gradis* Boheman, the cotton fleahopper, *Pseudatomoscelis seriatus* (Reuter), lygus bugs, *Lygus* spp., and the bollworm–*tobacco-budworm complex, Heliothis* spp. This research has focused on the development of insect-resistant cotton varieties, natural enemies, microbial pesticides, pheromone traps, and other control tactics with potential integration into a multifactorial pest management system.

The use of systems analysis and computer models as a means of unifying and guiding research and gaining understanding of system interactions and the consequences of applying these control tactics have also been a central part of this research. Although the value of this work entailing a systems approach remains to be fully tested and evaluated for pest management applications in cotton and other agricultural crops, benefits have already been realized. For example, it has greatly facilitated research, especially in identifying research gaps and priorities through modeling, sensitivity analysis, and validation. These benefits have accrued not only from research focusing on pest management in cotton but also on the management of pests of other agricultural crops where a similar approach has been taken (Huffaker and Croft, 1975).

The Cotton Ecosystem

The concept of the agricultural crop as an ecological system or agroecosystem is discussed in Chapter VI. Articles by Smith and van den Bosch (1967) and Smith and Reynolds (1972) present an exceptionally good discussion of how these systems may be mismanaged to aggravate pest problems or properly managed for more effective control of pest populations.

Smith and van den Bosch (1967) defined the agroecosystem as "a unit composed of the total complex of organisms in the crop area together with the overall conditioning environment as modified by the various agricultural, industrial, social, and recreational activities of man." They listed the major components of such a system as the crop plants, the soil substrate and its essential biota, the chemical and physical environment, an energy input from the sun and man. The agroecosystem, as Doutt (1964) pointed out, should be considered more as a man-manipulated system than as a natural one.

Smith and Reynolds (1972) describe the cotton agroecosystem as a complex biological system composed of many interlocking components, some man-controlled and some not, that affect pest population densities. Their concept of the cotton agroecosystem includes, in addition to the cotton fields in a given area, the association of fields of other agricultural crops in the area together with their marginal areas and often other intermixed areas such as woods, streams, and weedy or uncultivated areas. While we agree with this concept, we choose to use the term ''cotton ecosystem,'' and in the discussion to follow will delimit this system to one cotton field or to a group of associated cotton fields, i.e., without the associated noncotton agricultural crops and other external land areas.

Modeling the Cotton Ecosystem

There have been numerous definitions advanced to describe a model, depending on the model's use (description, prediction, etc.) and the form it takes (deterministic, stochastic, etc.), and it is rare when any two modelers reach a common definition. The simple definition used by Ruesink (1975) is adequate for our discussion: a model is an imitation and representation of the real world.* Thus, the ultimate goal in modeling the cotton ecosystem is to develop a computer model which realistically imitates and represents all occurrences that take place in a particular cotton field or group of fields at any given time from planting time through crop harvest. The desirability of such a model is apparent. For example, this model could be used in simulation studies to determine how the ecosystem can be manipulated, i.e., use of a particular combination of cotton plant variety, fertilizer, insect control practice, etc., to achieve optimal yield for the farmer. Although the value of a realistic simulation model is undisputed, development of a realistic model of the cotton ecosystem has not been achieved. And, in spite of certain claims, it is doubtful if such a model exists for any agricultural crop ecosystem.

Nevertheless, the recent assemblage of interdisciplinary research teams composed of entomologists, plant pathologists, agronomists, meteorologists, and members of other biological and physical disciplines has provided the basic blueprint from which the systems analyst has begun the ambitious effort of

* The reader is referred to two recent articles by Ruesink (1975, 1976) for a comprehensive discussion of the different types of models and recent developments in the general field of pest management modeling.

building a meaningful cotton ecosystem model. Initially, these teams provided the systems analyst with a list of the variables and causal pathways that were judged of potential importance in determining the function of this ecosystem, i.e., a conceptual model. Concurrently, they began to conduct laboratory and field experiments to provide data that the systems analyst could use to quantify and verify the importance of these variables, to formulate and test hypotheses, etc. We will now discuss some of the progress our research group in Texas has made toward developing an overall simulation model of the cotton ecosystem and submodels of certain of the system's components.

For purposes of simplicity, the cotton ecosystem will be divided conceptually into three levels: (1) the physiological level, (2) the population level, and (3) the economic or decision level. This simplistic ecosystem is inhabited by two primary groups of organisms: plants and insects (and, of course, many other organisms such as plant pathogens, spiders, etc.). Modeling activities within our research group have focused on all three levels, with work underway on both plants and insects at each of the first two levels.

Physiological level models are being developed as a basis for population models of the two principal key pests of cotton in Texas, boll weevil and cotton fleahopper, a braconid parasite, *Bracon mellitor* Say, of the boll weevil and the cotton plant. The population models are being developed as part of the economic and decision models. In addition, the population models are being used as a guide for interpretation of field experiments and will be used in future planning and analysis of additional field experiments. Economic and decision models are only in the planning state but are intended for later use in analyzing control strategies and in determining the types and quantity of data needed in making rational economic decisions in pest management.

A generalized flow chart for a simulation model of the simplistic cotton ecosystem described above is presented in Figure 1. This system is, of course, composed of several major components or subsystems involving the plant, local plant environment, plant growth, structure, and fruit setting characteristics. Coupled with the overall plant system and its local environment are the insect pest, parasite, and predator subsystems. In the diagram, connecting lines are drawn in to pinpoint the interactions between the various subsystems. The solid lines indicate that the subsystem interactions are reasonably well understood and that models have been developed to describe the interactions. The dotted lines represent those areas in which additional work needs to be done.

The plant's local environment model was developed by Lemon and Stewart (1969). Given the general environment surrounding the cotton crop, this model (SPAM) can be used to predict the local environment in the vicinity of any given plant. The plant growth model itself can be broken into several parts. The general plant growth model (LEAF) was developed by DeMichele and Sharpe (1973) and Sharpe and DeMichele (1974). This model, given the local

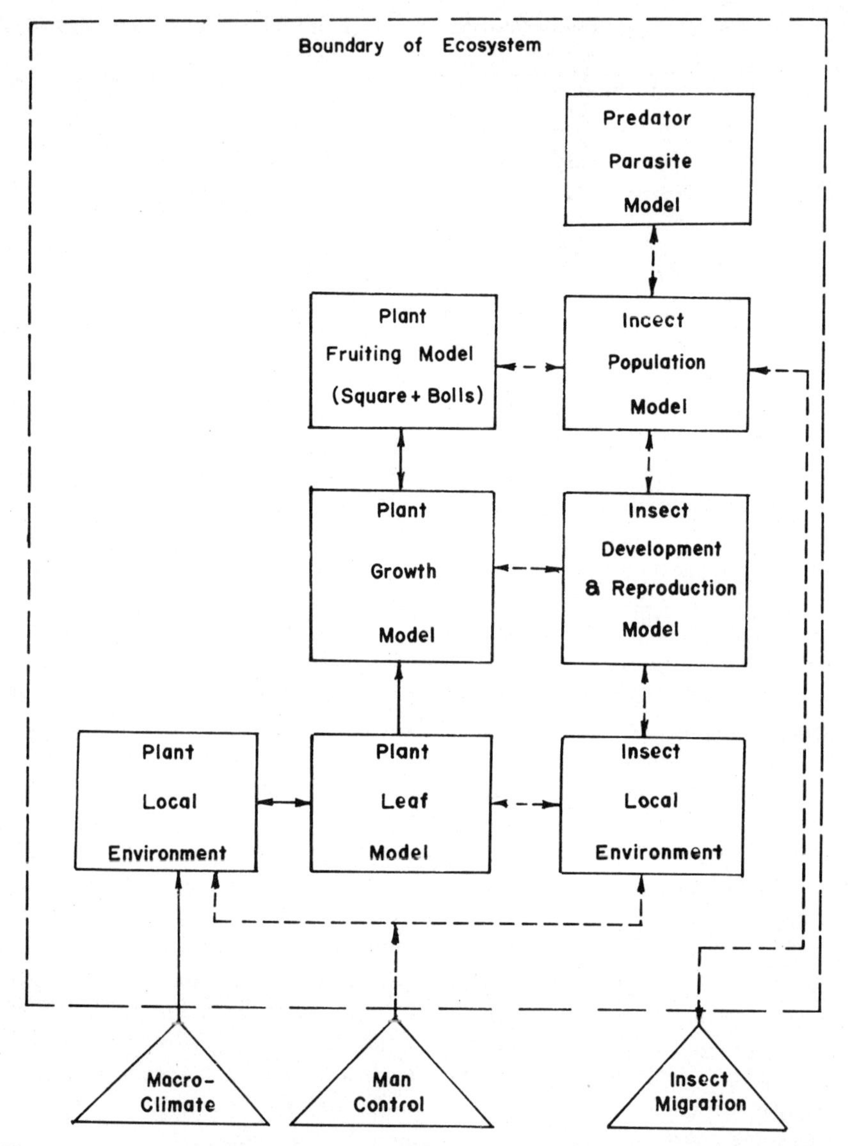

Figure 1. A generalized flow chart for a simulation model of cotton ecosystem (D. W. DeMichele, G. L. Curry, and P. Sharpe, unpublished).

plant environment via SPAM, can be used to predict the rate of photosynthesis, the rate of water usage, and the local environment of the plant leaf. The growth, location, and abscission of squares (flower buds) on the fruiting branches can be predicted with SIMCOT, a cotton plant model developed by

Baker and Hesketh (1970). Currently, we are in the process of coupling LEAF and SIMCOT.

Cotton Plant–Insect Pest Interactions

Several early twentieth-century cotton entomologists recognized the effect of the plant's growth stage and condition on pest infestations and how this effect could influence control decisions. It was recognized that the cotton plant produced many more squares than would develop into bolls, thus, it was considered unprofitable to attempt to protect all squares from insect attack. It also was recognized that insect damage to squares early in the season had a greater effect on yield than damage later in the season. Many modern day researchers also have stressed the usefulness of information on the plant's growth chacteristics in research, pest control decision making, and overall crop management (Reynolds *et al.*, 1975). This information has special importance in establishing realistic economic injury and threshold levels for various cotton pests and in determining the effect of different growth characteristics (as manifested by variety, moisture, fertility, etc. interactions) on the pest.

Our work with the boll weevil has revealed that population models cannot be developed for this insect without concurrently modeling the effect of its host plants. This only seems logical since the plant regulates the insect's temperature to a certain degree, its food supply and reproduction sites, and, hence, its developmental rates and fecundity.

The importance of the cotton plant on the boll weevil's reproduction rate can be seen in Figures 2 and 3. Using a population model (Sharpe *et al.*, 1975) that incorporates data on fecundity, mortality, and developmental rates of this pest, a simulation of a natural population was attempted with and without the cotton plant's interface. In simulating without the plant's interface, we assumed that the overwintering population (FO) began reproducing in the spring as soon as the first squares reached a size (≥0.15 in. diam) considered necessary for oviposition by females. It was assumed that oviposition by females began independent of the per acre density of squares this size. Simulation with the plant's interface assumed that egg oviposition by females began only after the density of squares (≥0.25 in. diam) reached 10,000 per acre. The value of 10,000 per acre was used since field data from several studies indicated, for reasons not known, that this density correlates closely with the onset of egg oviposition by FO females.

Very poor population simulation was achieved with two different sets of field data when the plant interface was deleted from the model (Figures 2 and 3). The *F* population was greatly overestimated in both cases. However, as evident from the figures, inclusion of the plant's interface, i.e., adjustment for

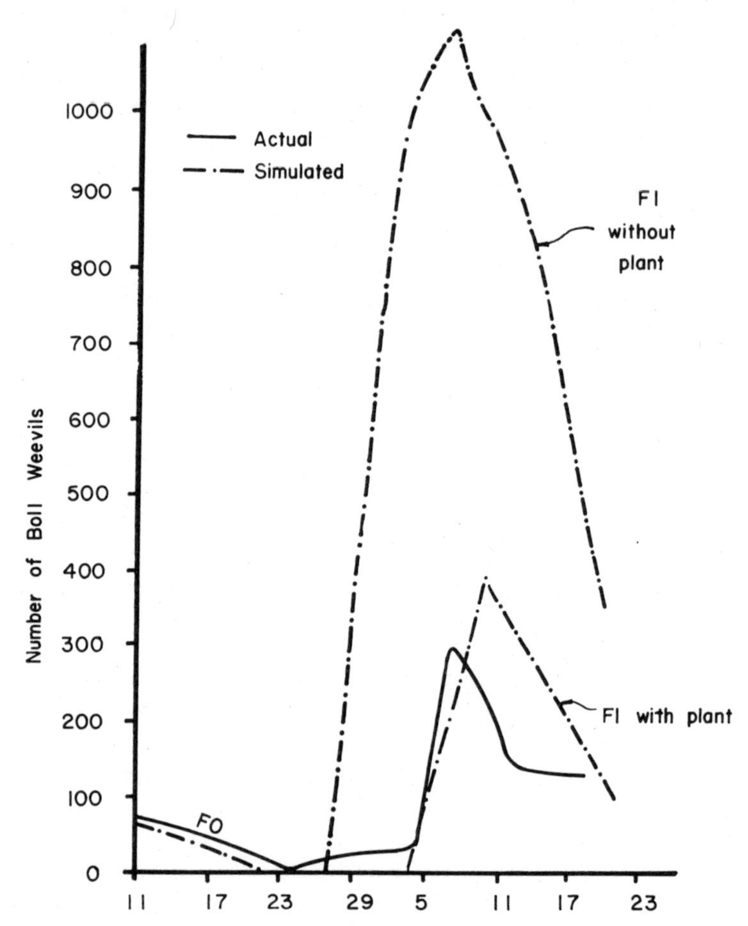

Figure 2. Effect of the cotton plant's squaring patterns on boll weevil reproduction. Refer to the text for explanation (actual data provided by J. K. Walker, Jr.; simulation model developed by G. L. Curry, D. W. DeMichele, and P. Sharpe).

square density (10,000 per acre) thought necessary for initiation of reproduction, gave much better simulation, although less than we eventually hope to achieve. Why is 10,000 a seemingly magic number for the boll weevil? Could it be that the corresponding concentration of certain plant attractants present at square density 10,000 per acre is required to incite egg laying? Or could it simply be that this density of squares must be reached before the female boll weevils can adequately seek out plant food necessary for development of the reproductive system?

These are the kinds of questions that our modeling venture has evoked and that must be answered before a meaningful biological model can ever be developed to *explain* the boll weevil–cotton plant interface. Because of the power and flexibility of simulation techniques, it is usually a simple matter to construct a model that will match a set of observations (Paulik and Greenough, 1966). It was a very simple matter with the model just discussed to make appropriate adjustments as required to match the field data with the model output data. The production of such realistic appearing output is, however, no assurance that the model provides a valid representation of a particular biological system (Paulik and Greenough, 1966).

Now we will give an example of a model built solely from physical and biological concepts and one that has not needed the help of "adjustments" to

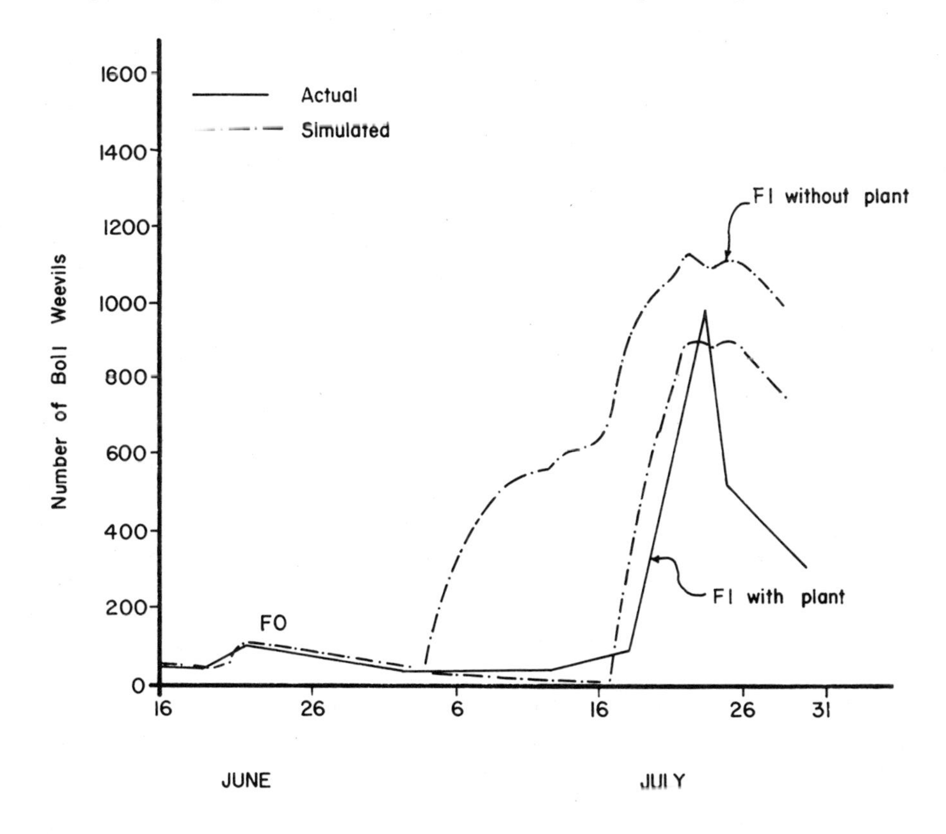

Figure 3. Effect of the cotton plant's squaring patterns on boll weevil reproduction. Refer to the text for explanation (actual data provided by J. K. Walker, Jr.; simulation model developed by G. L. Curry, D. W. DeMichele, and P. Sharpe).

produce accurate results. While building a model from concepts is more difficult than building one using "adjustment" factors, we believe strongly that it is the only model capable of standing the test of time in any pest management modeling effort.

The condition of the cotton squares and bolls sometimes affects not only the development and fecundity of certain pests but also the survival of the immature stages that develop inside the fruiting structures. To determine the effect of square condition on survival of immature boll weevils, a square drying model was developed to describe the effects of the environment on abscised squares (DeMichele *et al.*, 1975; Curry *et al.*, 1975). The basic assumptions in development of the model were

1. The square could be approximated by an ellipsoid of revolution.
2. The square dries from the outside inward.
3. The square tissue is homogeneous in its diffusion properties.

The model was used to predict the progressive loss of water from cotton squares of three different sizes which had been abscised from the plants. We assumed the water loss was proportional to loss of viable mass, i.e., potential insect food. Hence, the model actually predicts the amount of food reserve remaining in a square exposed to a given environmental condition for a certain period of time.

The model predicts very accurately the progressive loss of water from squares of different sizes (Figure 4). If water loss is indeed proportional to loss of viable mass, then this model should be extremely useful in predicting the probability of the immature weevil's maturation within squares exposed to different environments.

Using the same mathematical model, simulations were run to mimic square drying under conditions of varied temperature, humidity, and sunlight (Figure 5). Simulations were run for 17 days to approximate the length of time required for the immature boll weevil to develop to the pupal stage. It was assumed that square drying was significant only during a 12-hr daylight period of each of the 17 days.

These simulations disclosed the extreme importance of square size in affecting boll weevil survival under the different microenvironmental conditions (Figure 5). It is apparent that immature weevils residing in squares sheltered from sunlight, i.e., shaded by the plant's canopy, would have a greater chance of surviving to adulthood than immature weevils in exposed squares. The variables of temperature, humidity, and square size further affect the chance of survival.

Presently, we do not know the dietary requirement (viable mass) for immature weevil development. Hence, without this input the model cannot be ex-

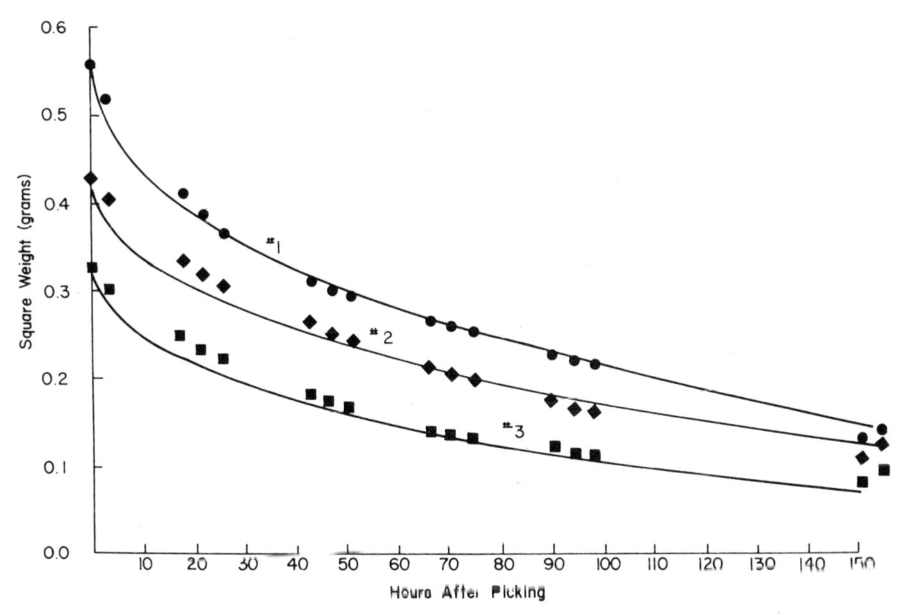

Figure 4. Weight loss of abscised cotton squares of three size classes (#1, #2, and #3) under conditions of constant temperature (25–27°C) and relative humidity (41–47%) (actual data provided by C. S. Barfield; model developed by D. W. DeMichele, P. Sharpe, and G. L. Curry).

pected to produce highly realistic simulations. Once the dietary need has been determined, we visualize numerous applications with this model. For instance, simulation from the model could be used as a guideline in determining the value of certain cultural practices (irrigation, plant row spacing patterns, etc.) in creating an environment which would induce highest levels of mortality in the immature weevil population.

Cotton Crop Growth Models

The focus of any crop protection program must be the crop, and thus a crop growth and development model should be the central feature of any systems approach to crop protection (Ruesink, 1975). The utility of such a model as a research tool is obvious. A realistic crop growth model would permit rapid computer screening of insect-resisting characters, examination of pesticidal effect on fruiting characteristics, etc. When properly coupled with other subsys-

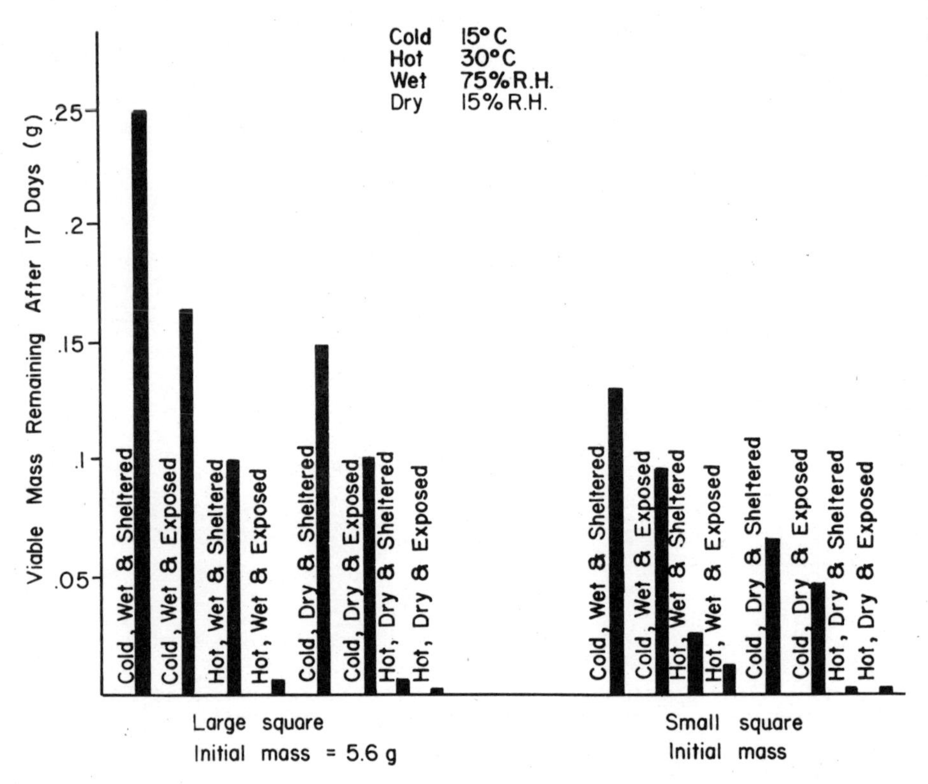

Figure 5. Effect of microenvironment on abscised cotton squares (from unpublished data of D. W. DeMichele, P. Sharpe, and G. L. Curry).

tem models, it would facilitate simulation of total production systems including economic and environmental consequences and offer great utility in many phases of farm management, marketing, and environmental considerations.

There are several cotton plant modeling projects currently underway at several federal and state laboratories (Wanjura *et al.*, 1975; Weaver *et al.*, 1975; Gutierrez *et al.*, 1975; Huffaker and Croft, 1975). The pioneering work in cotton plant modeling was begun in the mid-1960s by personnel of ARS–USDA in Mississippi who developed the single-plant model, SIMCOT II (McKinion et al., 1974). This model has been used with varying success in simulation studies of growth and development. Gutierrez *et al.* (1975) successfully modified SIMCOT II to allow simulation of a field population of cotton plants grown under California conditions. Field validation of the California version, SIMCTO-UC, indicates that this model may provide harvest yield simulations with about 90% accuracy. This level of accuracy, if achieved consistently, would be adequate in many phases of pest management research and implementation.

We have not been able to achieve accurate simulation results with SIMCOT II in Texas. We envision a major revision in this model before it can provide consistently accurate simulations in most growing regions of Texas, where the environmental conditions are much less stable than in California.

Dispersal and Pheromone Drift Models

Many cotton insects are powerful flyers, and this characteristic often poses difficulty in the design and implementation of pest management programs. Years of experience in controlling the boll weevil, for example, have revealed that the pest can best be managed when approached on a large-district basis as opposed to an individual farm or field basis. In fact, a farmer can eliminate an economic-level weevil infestation with insecticides one day only to be flooded with a migrant population the next if a neighbor did not take the same action.

A question often raised when establishing pest management districts (communitywide control programs) for the boll weevil is that of the isolation required from noncontrol districts to buffer the effect of reinfestation dispersal? This has not been adequately determined due to the logistics problem in mark-release-recapture and other similar research efforts to quantify the distance of population dispersal. We feel that a computer model of this insect's dispersal patterns offers great potential in circumventing many of these research obstacles.

A mathematical model has recently been constructed, although not yet validated, to predict the emigratory patterns of a population of boll weevils from a point source. Figure 6 shows the predicted density distribution patterns of a population of weevils emigrating randomly, in the absence of wind, from a point represented by the peak at the center of the computer plot. The theoretical effect of a strong wind on this distribution is portrayed in Figure 7. Once validated under actual field conditions, this model could serve as a guide for assigning boundaries to pest management districts such as are not being formed in many parts of the southeastern United States.

Modeling may also have practical utility in research on cotton insect pheromones. Presently, a model is being developed to help clarify the probable effects of inversion layers on the dispersion rate of the boll weevil's synthetic pheromone (grandlure) and the effects of wind and temperature on emission rates. It would be extremely difficult, if not impossible, to determine these effects experimentally under field conditions. The preliminary model, however, has revealed some insight into the possible effects.

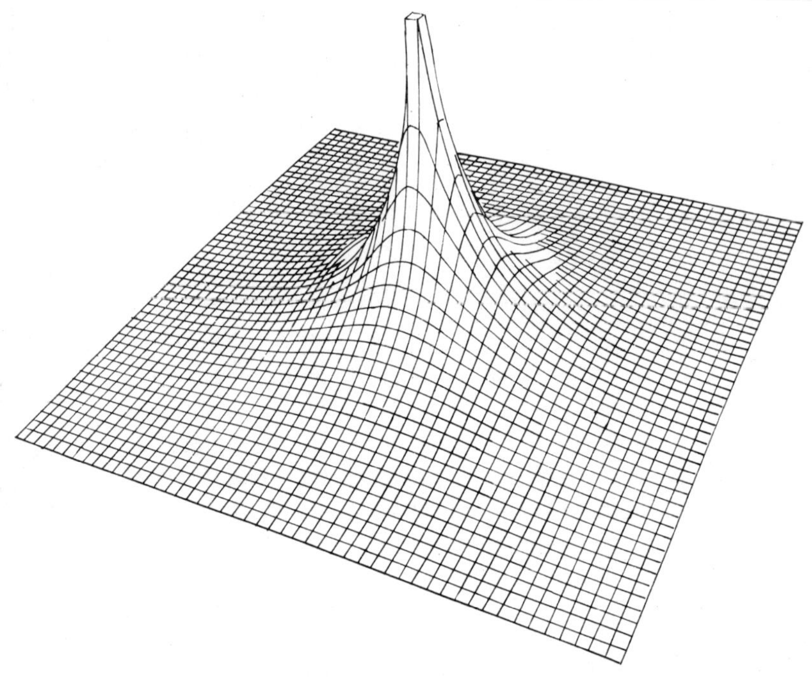

Figure 6. Theoretical emigratory patterns of a population of boll weevils when not influenced by wind or pheromone attraction. Vertical axis represents pest density (D. DuBose, unpublished).

The theoretical dispersion of pheromone from eight point sources (traps) in a hypothetical cotton field is illustrated in Figures 8 and 9. The first figure portrays the expected dispersion at midday under conditions of high wind velocity and no cloud cover. The second figure portrays an altered effect which might be expected to result from a night temperature inversion layer and a much lower wind velocity.

Modeling efforts on boll weevil pheromone trap response are also underway in Mississippi (Jones *et al.,* 1975), and there are several major projects focusing on migration of the bollworm–tobacco budworm complex (Stinner *et al.,* 1974; Hartstack *et al.,* 1975).

Uses of Models and Systems Analysis in Decision Making

The typical cotton ecosystem is controlled primarily by the driving force of the environment and the agronomic practices of the farmer. Major components of the system, as illustrated in Figure 1., were discussed earlier. The important weather factors are, of course, sunlight, humidity, air temperature, wind speed, rainfall, and photoperiod. Of these, only photoperiod can be predicted consistently with high accuracy. Coupled with the tremendous biological variations in individuals of the crop and animal populations are the variations in microclimate, nutrient supply, pest density, and agricultural practices within the field. The effects of environment on growth, development, reproduction, disease resistance, and yield are not clearly understood with the exception of a few specified conditions. Other than under these specified conditions, the interrelations are essentially unknown.

It is generally accepted that there are no random events in nature other than perhaps at the level of individual nucleons and electrons. The *apparent* randomness stems from a lack of detailed knowledge about the mechanics of the system. If we know nothing about a system, the response must be treated as a completely random expectation. As we begin to acquire data and develop theories, the outcomes, although partially random, are characterized by a nonuniform distribution function. A system whose outcome is predictable is one where all mechanisms and inputs to the system are known. The distribution

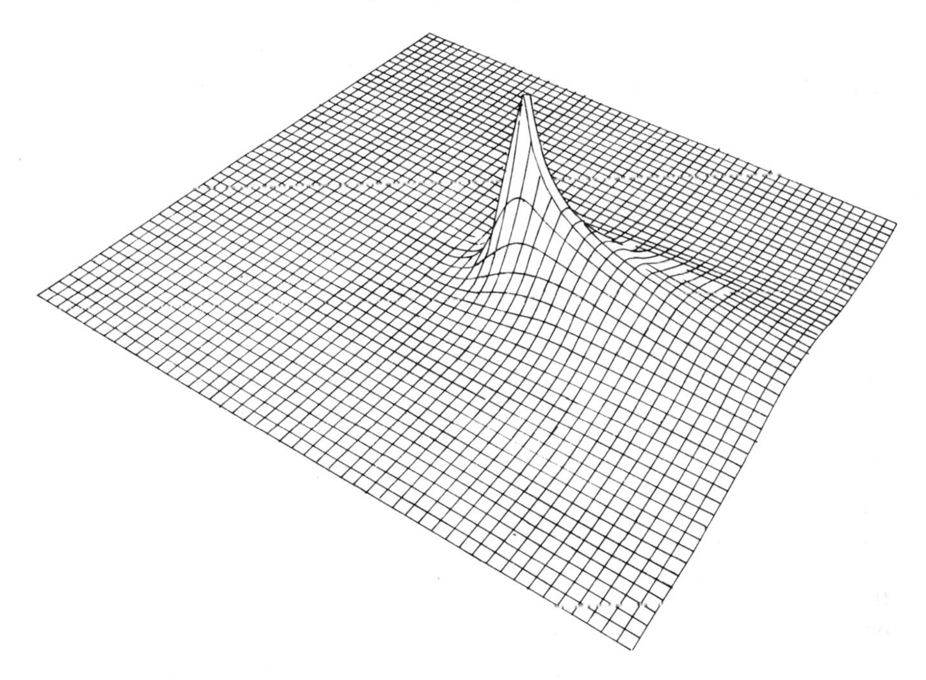

Figure 7. Theoretical emigratory patterns of a population of boll weevils as influenced by wind. Vertical axis represents pest density (D. DuBose, unpublished).

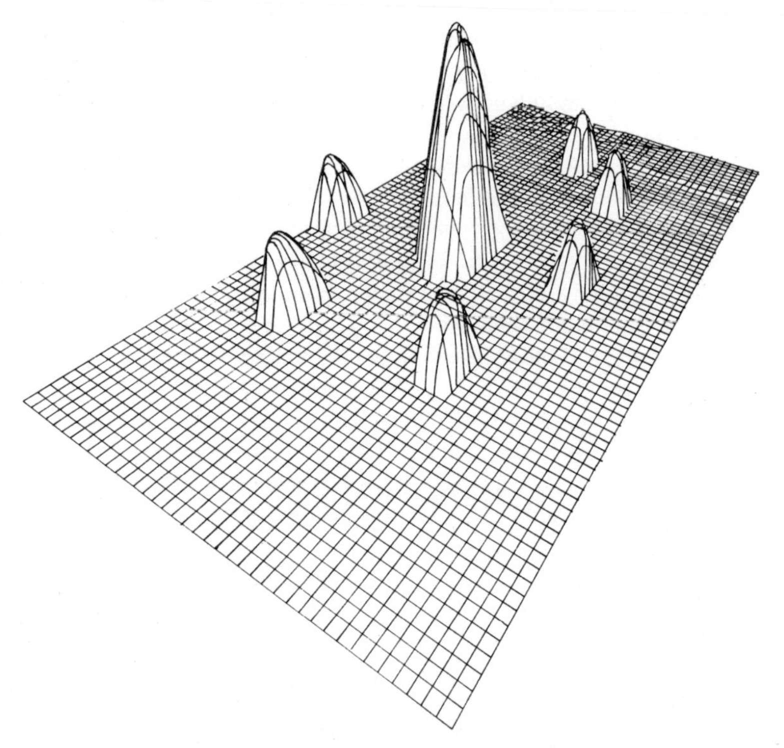

Figure 8. Theoretical midday dispersion of boll weevil pheromone (grandlure) from eight point sources (traps) in an imaginary cotton field exposed to conditions of high wind velocity. Vertical axis represents pheromone concentration at ground level (D. DuBose and D. W. DeMichele, unpublished).

function becomes a delta function where the probability of all but one of the outcomes is zero. Figure 10 shows the modification in outcome predictability as we become more and more knowledgeable about a system. Due to our present lack of knowledge the cotton ecosystem must be viewed as a stochastic or random system and, therefore, the response to any specified agronomic or pest management practice must be considered as a random variable. Some of the models currently being developed will help to sharpen the distribution of the random variable, but as should be evident from earlier discussions, we are far from a delta function distribution in our efforts to model the overall cotton ecosystem, even though we are approaching this distribution with certain of its subsystems.

The Decision Process. Most decision processes requiring detailed analysis are characterized by a sequence of decisions and chance occurrences. That is, for a given decision, a whole set of outcomes is possible. After realizing a specific outcome, the decision-maker will be faced with new decisions which are determined by an initial decision and by the outcome of that decision. A second decision is made, and again chance becomes a dominant factor in the resultant outcome.

The problems confronting a cotton farmer, faced with making a decision of applying a pesticide, for example, can be schematically illustrated in a hypothetical decision tree that outlines alternative pesticide application policies (Figure 11). There are two types of branch nodes shown in this diagram. The square nodes represent points where decisions are to be made, and the circular nodes represent the outcome of chance. It has been argued that, since we may lack a complete understanding of a given system (e.g., cotton ecosystem), the specific outcome or response of this system (to natural forces or to actions imposed by the farmer) must be regarded as only a random variable. The best we could hope for is a better prediction probability. In the example shown in Figure 11, the farmer initially makes a pesticide application decision at node #1.

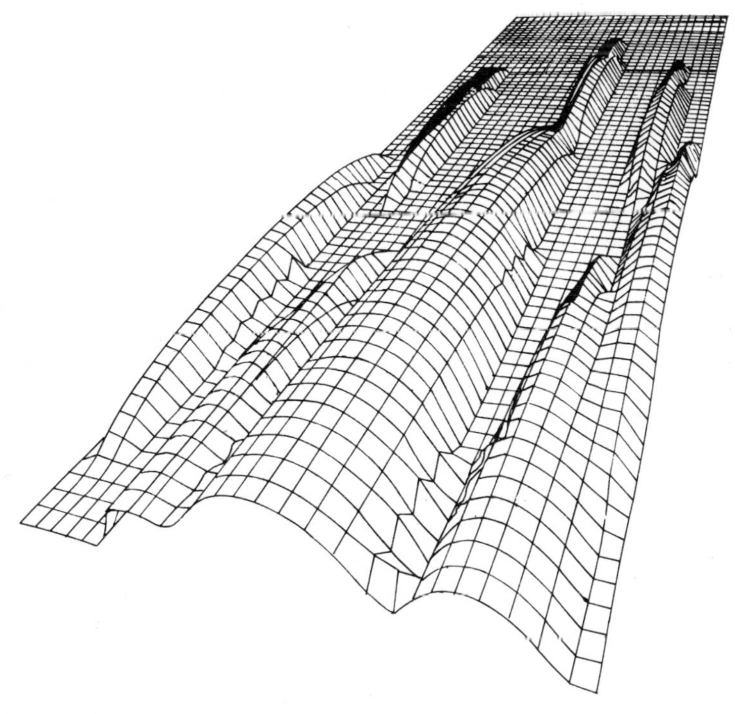

Figure 9. The theoretical effect of a night temperature inversion and a reduced wind velocity on dispersion of pheromone from the eight point sources shown in Figure 8 (D. DuBose and D. W. DeMichele, unpublished).

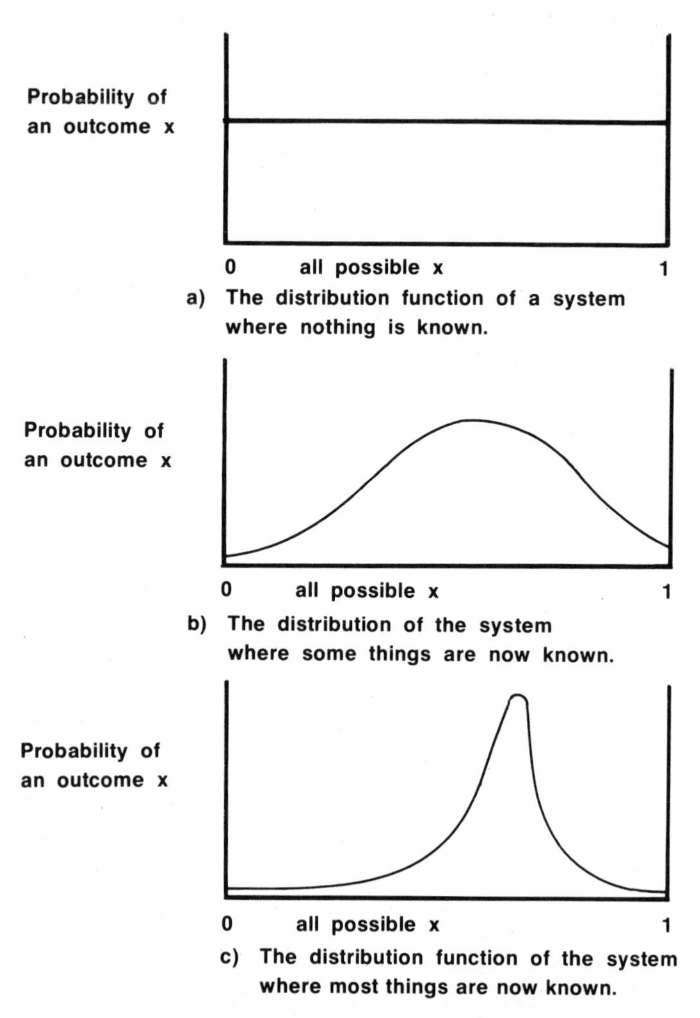

Figure 10. Process of outcome predictability based on amount of knowledge of a system (D. W. DeMichele, P. Sharpe, and G. L. Curry, unpublished).

If he decides not to apply any pesticide during the season, then he will be subject to chance (C_4) and can expect to get a return of R_7 or R_8. The chance values of C_1 must be determined by some type of plant–insect ecosystem production model. If the farmer decides to apply a pesticide, then he must decide when and how much. If he decides on an initial application, he is subject to a probability (C_1) of an infestation, and if this infestation occurs, he must take decision #2, and so forth. At the tip of each branch is a payoff or return, R_1, representing a given utility to the farmer. The values of R_1 are determined by

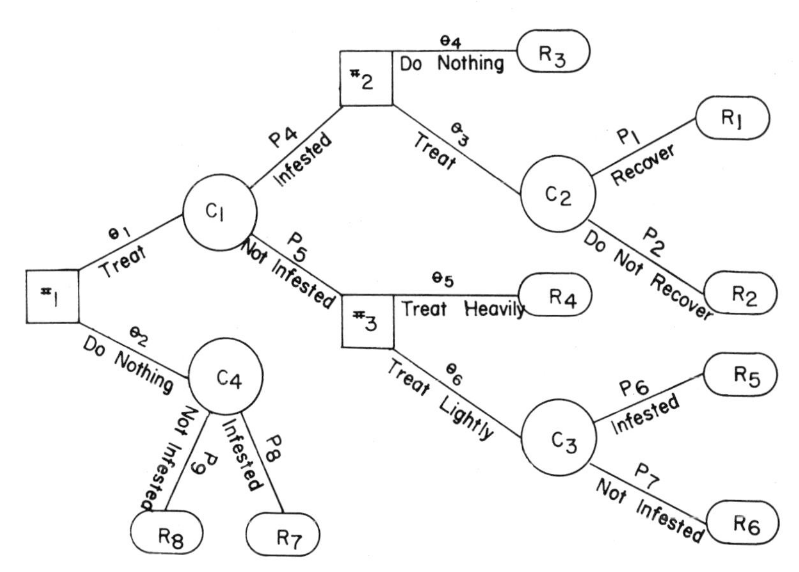

Figure 11. Typical decision tree.

this individual's own judgment. The objective of the decision analysis is to determine the best strategy for each possible set of chance outcomes. The relative value of any decision is determined by the farmer's utility with respect to all possible returns. It is well-known that most people are risk-adverse; therefore, to lessen these risks, decision makers need to be knowledgeable not only of the expected value of the return for any sequence of decisions but also of the variance in the value of the possible outcomes. A farmer may resist following a course of action, e.g., pesticide treatment, which has a high expected return but also a possibility of a significant loss. Therefore, for greatest utility in their application centered around pest control, models must yield not only an expected outcome but also the probability distribution of all possible outcomes. The "optimal" decision sequence that we are seeking is based, in part, on the response of the model and, in part, on the utility of the farmer. Newton and Leuschner (1975) provide a further discussion of decision making under risk in pest management programs.

Recognizing the great variability in the cotton ecosystem and the needs of the farmer, we must seek to build models that yield not only the established probability of a whole spectrum of possible outcomes but also an estimate of the accuracy (realism) of the answers that the models supply. This is a most ambitious undertaking that will require close cooperation of several disciplines engaged in agricultural research and back-up support from extension specialists, but is a goal that we all should seek.

Conclusions

The past decade has produced many examples which should serve as proof of the many advantages in controlling agricultural crop pests via integrated management strategies centered around ecological principles. As evident from examples cited in the article by Huffaker and Croft (1975) and from the many examples discussed in this book, integrated pest management systems have been developed for pests of the major agricultural crops in the United States. If these systems now available could be properly implemented throughout the entire range in this country where applicable, they would produce incalculable dividends to American agriculture. Perhaps equally, if not more important, these systems portend to lessen greatly the major problems of environmental pollution and human health hazards inherent to the many unilateral insecticide-based pest control programs that still dominate today. Various estimates suggest that the adoption of the currently available integrated insect pest management systems would permit a 40–50% reduction in the use of the more environmentally polluting insecticides within a 5-year period. A reduction of 70–80% is envisioned in the next 10 years. Many pest management experts believe that these reductions are possible with no resulting sacrifice in crop yield or farmer profit.

Opportunities for implementing sound integrated pest management systems look especially good in cotton, which historically has been a major consumer of pesticides in the United States (Reynolds *et al.*, 1975). However, in spite of the fact that technology is now available for the successful application of these systems to a large segment of the cotton belt, this technology is being adopted very slowly at the farmer level. And we seriously doubt that the present system of farm management will be conducive to either the testing or the adoption of much of this technology.

This disparity in development of new pest control technology and its adoption by farmers is obviously more than a simple technological problem. Rather, it is what Luckman (1975) describes as an obstacle resulting from the "attitude of people"—growers, industry personnel, and most important of all, applied research entomologists and extension personnel. If the attitudes of these key people are not in accord, the development of new pest management technology offers no real utility to society because it likely will not be adopted successfully by the farmer. In short, crop pest specialists at all levels of research, extension, and agribusiness activity must be committed to the implementation of pest management programs based on new technology if the concept is to fulfill its potential for crop protection.

The recent trend toward increased emphasis on pest management "action" programs, the expressed attitudes of the many authors herein and elsewhere and the prevailing attitudes of the American farmer and society as a whole would indicate, however, that pest management has high potential utility with value to all. However, as we have tried to point out here and as others have pointed out elsewhere, utility theory is founded on the concept that individual preferences reflect the utility of the thing or consequence being selected. Thus, a preferred consequence has a greater utility value than a less preferred consequence (Newton and Leuschner, 1975). Thus, a cotton farmer may prefer, in philosophy, a new pest management tactic over a systematic insecticide application, but in practice he may resist acceptance of the former control tactic if he is uncertain of its associated risks. In turn, if the pest control specialist is uncertain of these risks, he will resist recommending the "preferred" tactic.

The so-called systems approach which embraces the methodologies of systems analysis, modeling, and computer hardware offers great potential in providing both the farmer and the pest control specialist a decision-making capability whereby the risks are known for the preferred and alternative courses of action. We believe that this should be the ultimate goal of any systems approach effort centered around pest management. And we believe that farmers will buy emerging pest management programs as a preferred strategy once they are convinced that the benefits of these programs outweigh their risks. In anticipation that the technology for delivery is forthcoming in many cotton growing regions, we would encourage research and extension specialists to begin now to explore the use of on-line computer delivery systems at the farm level. Computer delivery systems such as now in use in apple pest management programs in Michigan (Croft, 1975) have great immediate potential in cotton pest management programs, in spite of our lack of realistic crop–insect management models in cotton. Algorithms have been developed for these systems which facilitate instant dissemination of extension information on weather patterns, crop variety performance, etc., upon command by the farmer or his extension specialist. These systems portend to greatly strengthen the farmer's confidence in the specialist's service and at the same time prepare them both for forthcoming developments in more sophisticated computer-based pest management technology.

In summary, the terms "systems approach," "modeling," and "systems analysis," never heard of by many entomologists before the beginning of this decade, have become household words within the profession. We believe that the science of modeling and systems analysis, presently in its infancy in pest management, will serve a prerequisite role in the unraveling of some of our most difficult pest control problems of the future.

ACKNOWLEDGMENTS

This work was supported in part by the National Science Foundation and the Environmental Protection Agency, through a grant (NSF GB-34718) to the University of California. The work was in cooperation with ARS-USDA and funded in part under USDA Cooperative Agreement No. 12-14-100-11,194(33). The findings, opinions, and recommendations expressed herein are those of the authors and not necessarily those of the University of California, the National Science Foundation, the Environmental Protection Agency, or ARS–USDA.

Special thanks are due Drs. G. L. Curry and P. J. H. Sharpe for providing much of the data presented herein and for other valuable assistance.

Literature Cited

Adkisson, P. L., 1973*a,* The integrated control of the insect pests of cotton, *in:* Proceedings of the Tall Timbers Conference on Ecological Animal Control by Habitat Management, No. 4. pp. 175–188.

Adkisson, P. L., 1973*b*. The principles, strategies and tactics of pest control in cotton, *in: Insects: Studies in Population Management* (P. W. Geier, L. R. Clark, D. J. Anderson, and H. A. Nix, eds.), Ecology Society of Australia (Memoirs 1), Canberra. p. 274–283.

Baker, D. N., and Hesketh, J. D., 1970, SIMCOT. Annual Report of the Boll Weevil Research Laboratory, ARS–USDA, State College, Miss.

Bottrell, D. G., 1973, Development of principles for managing insect populations in the cotton ecosystem: *in:* Texas, Proceedings of the Beltwide Cotton Production Research Conference, pp. 82–84.

Croft, B. A., 1975, Tree fruit pest management, *in: Introduction to Pest Management* (R. L. Metcalf and W. Luckmann, eds.), John Wiley and Sons, New York, pp. 471–507.

Curry, G. L., Sharpe, P. J. H., DeMichele, D. W., and Bottrell, D. G., 1975, Cotton square drying. 2. Implications for immature boll weevil survival, *Environ. Entomol.* (Accepted).

DeMichele, D. W., Curry, G. L., Sharpe, P. J. H., and Barfield, C. S., 1975, Cotton square drying. 1. A theoretical model, *Environ. Entomol.* (Accepted).

DeMichele, D. W., and Sharpe, P. J. H., 1973, An analysis of the mechanics of guard cell motion, *J. Theor. Biol.* **41:**77–96.

Doutt, R. L., 1964, Ecological considerations in chemical control: Implications to non-target invertebrates, *Bull. Entomol. Soc. Am.* **10:**67–88.

Gutierrez, A. P., Falcon, L. A., Loew, W., Leipzig, P. A., and van den Borsch, R., 1975, An analysis of cotton production in California: A model for Acala cotton and the effects of defoliators on its yields, *Environ. Entomol.* **4:**125–136.

Harris, F. A., 1973, Development of principles for managing insect populations in the cotton ecosystem as related to Mississippi, *in:* Proceedings of the Beltwide Cotton Production Research Conference, pp. 86–88.

Hartstack, A. W., Witz, J. A., and Ridgway, R. L., 1975, Suggested applications of a dynamic Heliothis model (MOTHZV-1) in pest management decision making, *in:* Proceedings of the Beltwide Cotton Production Research Conference, pp. 118–122.

Huffaker, C. B., and Croft, B. A., 1975, Integrated pest management in the USA-progress and promise, *Environ. Health Persp.* (In press).

Huffaker, C. B., and Smith, R. F., 1973, The IBP program on the strategies and tactics of pest management, *in:* Proceedings of the Tall Timbers Conference on Ecological Animal Control by Habitat Management, No. 4, pp. 219–236.

Jenkins, G. M., 1969, The systems approach, *J. Syst. Eng.* **1:**1–17.

Jones, J. W., Thompson, A. C., and McKinion, J. M., 1975, Developing a computer model with various control methods for eradication of boll weevils, *in:* Proceedings of the Beltwide Cotton Production Research Conference, p. 118.

Lemon, E. R., and Stewart, D. W., 1969, A simulation of net photosynthesis of field corn, Technical Report ECOM-68, Ithaca, N.Y.

Luckmann, W., 1975, Pest management and the future, *in: Introduction to Pest Management* (R. L. Metcalf and W. Luckmann, eds.), John Wiley and Sons, New York, pp. 567–570.

McKinion, J. M., Jones, J. W., and Hesketh, J. D., 1974, Analysis of SIMCOT: Photosynthesis and growth, *in:* Proceedings of the Beltwide Cotton Production Research Conferences, pp. 118–125.

Newton, C. M., and Leuschner, W. A., 1975, Recognition of risk and utility in pest management decisions, *Bull. Entomol. Soc. Am.* **21:**169–172.

Paulik, G. J., and Greenough, J. W., Jr. 1966, Management analysis for a salmon resource system, *in: Systems Analysis in Ecology* (K. E. F. Watt, ed.), Academic Press, New York, 273 p.

Phillips, J. R., 1973, The NSF/IBP cotton research program in Arkansas—A first year evaluation, *in:* Proceedings of the Beltwide Cotton Production Research Conference, pp. 84–86.

Ranney, C. D., 1973, The 1973 national cotton research task force report. U.S. Department of Agriculture, Washington, D.C., 153 p.

Reynolds, H. T., 1973, Development of principles for managing insect populations in the cotton ecosystem—California, *in:* Proceedings in the Beltwide Cotton Production Research Conference, pp. 81–82.

Reynolds, H. T., Adkisson, P. L., and Smith, R. F., 1975, Cotton insect pest management, *in: Introduction to Pest Management* (R. L. Metcalf and W. Luckmann, eds.), John Wiley and Sons, New York, pp. 379–443.

Ruesink, W. G., 1975, Analysis and modeling in pest management, *in: Introduction to Pest Management* (R. L. Metcalf and W. Luckmann, eds.), John Wiley and Sons, New York. pp. 353–376.

Ruesink, W. G., 1976, Status of the systems approach to pest management, *Ann. Rev. Entomol.* (In press).

Sharpe, P. J. H., Curry, G. L., and DeMichele, D. W., 1975, A model of boll weevil growth and development, unpublished report.

Sharpe, P. J. H., and DeMichele, D. W., 1974, A morphological and physiological model of the leaf, unpublished report.

Smith, R. F., Huffaker, C. B., Adkisson, P. L., and Newsom, L. D., 1974, Progress achieved in the implementation of integrated control projects in the USA and tropical countries, OEPP/EPPO Bull. **4:**221–239.

Smith, R. F., and Falcon, L. A., 1973, Insect control for cotton in California, *Cotton Grow. Rev.* **50:**15–27.

Smith, R. F., and Huffaker, C. B., 1973, Integrated control strategy in the United States and its practical implementation, *OEPP/EPPO Bull.* **3:**31–49.

Smith, R. F., and Reynolds, H. T., 1972, Effects of manipulation of cotton agroecosystems on insect pest populations, *in: The Careless Technology—Ecology and International Development* (M. T. Farvar and J. P. Milton, eds.), Natural History Press, Garden City, N.Y., pp. 373–406.

Smith, R. F., and van den Bosch, R., 1967, Integrated control, *in: Pest Control: Biological, Physical, and Selected Chemical Methods* (W. W. Kilgore and R. L. Doutt, eds.), Academic Press, New York, pp. 295–340.

Stinner, R. E., Bradley, J. R., and Rabb, R. L., 1974, Population dynamics of *Heliothis zea* (Boddie) and *H. virescens* (F). in North Carolina: A simulation model, *Environ. Entomol.* **3:**163–168.

Twenty-seventh Annual Conference Report on Cotton Insect Research and Control, 1974 (ARS–*USDA* in cooperation with 13 cotton-growing states).

van den Bosch, R., Falcon, L. A., Gonzales, D., Hagen, K. S., Leigh, T. F., and Sterm, V. M., 1971, The developing program of integrated control of cotton pests in California, *in: Biological Control* (C. B. Huffaker, ed.), Plenum Press, New York, pp. 377–394.

Wanjura, D. F., Colwick, R. F., and Jones, J. W., 1975, Status of cotton-production-system modeling in regional research project S-69 *in:* Proceedings of the Beltwide Cotton Production Research Conference, pp. 1959–161.

Watt, K. E. F., 1966, The nature of systems analysis, *in: Systems Analysis in Ecology* (K. E. F. Watt, ed.), Academic Press, New York, pp. 1–14.

Weaver, R. E. C., Law, V. J., and Bailey, R. V., 1975, Bases for the use of single plant models in field studies, *in: Proceedings of the Beltwide Cotton Production Research Conference, pp. 168*–170.

Witz, J. A., 1973, Integration of systems science methodology and scientific research, *Agr. Sci. Rev.* **11:**37–48.

IX

Pest Management on Deciduous Fruits: Multidisciplinary Aspects

S. C. Hoyt and J. D. Gilpatrick

From the time the first broad-spectrum, synthetic pesticides were developed and used on deciduous fruits in the late 1940s until the early 1960s, pest control was treated as an isolated aspect of fruit production in many areas. Extensive chemical control programs utilizing insecticides, fungicides, and acaricides were developed. Frequently, the control of one pest would aggravate problems with a second pest, so additional chemicals were used to control the second pest. Orchard sanitation and cultural practices which might have aided in pest control were largely ignored.

In the past decade much greater emphasis has been placed on the regulation of pests through several tactics rather than total reliance on chemicals. It has been realized that the management of pests on deciduous fruits is not an isolated function, but is a portion of the overall management of the crop. Broadly, crop management can be broken down into two major areas: (1) pest management and (2) quality and yield management. In the area of pest management there are numerous species of insects and mites, bacterial, fungal, and viral diseases, weeds, and vertebrates which must be considered. Quality and yield management is concerned with the nutritional state of the tree, the control of tree growth, maintenance of good fruiting wood, fruit numbers and spacing, and the prevention of damage by pests. The factors influencing these two crop

---

S. C. HOYT · Tree Fruit Research Center, Washington Agricultural Experiment Station, Wenatchee, Washington 98801. J. D. GILPATRICK · New York Agricultural Experiment Station, Geneva, New York 14456.

management areas are highly interrelated and changes in methods in one area can have significant effects on the other.

In Figure 1 we have attempted to show the complexity of the interrelationships of the factors which must be managed in orchards, the cultural practices employed, and some external inputs which affect management. Within each factor to be managed there are relationships which are important. Examples of these are intra- and interspecific competition, the balance of various nutrients in relation to each other, or the relationship of fruit size to fruit numbers. There are also many critical relationships among the factors, such as the effect of nutrient deficiencies on pest populations, tree growth, and fruit development, or the influence of the cover crop on insect, rodent, or other pest densities. The external inputs of climate, weather, and the use of the land surrounding the orchard may have a direct effect on the factors to be managed. Finally, it is difficult to single out cultural practices which do not interact with one another or with the factors to be managed. In essence the entire ecosystem will respond to virtually all operations performed in the orchard.

Traditionally, the study of crop production has been conducted by research workers in the discipline most directly involved, with little or no input from other disciplinary areas even though interactions occurred. Usually the solution to a particular problem was treated as an event isolated from other problems on the crop. This has occurred even within disciplines. As we become more deeply involved in pest management, it becomes obvious that this type of approach is inadequate and that the interrelationships must be examined. Only in this way can the optimum practices for crop management be determined.

Most deciduous fruits are grown in the humid temperate areas of the world, but a significant percentage is also produced under semiarid conditions.

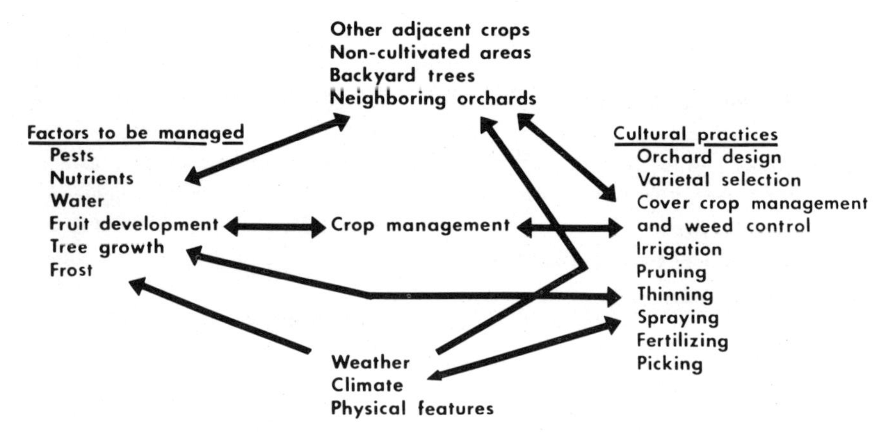

Figure 1. Interrelationships among factors involved in orchard management.

Pest management in humid areas is more complex because of the greater number of pest species which must be controlled. Most horticultural practices are similar in humid and arid areas, but major differences occur in cover crop and water management and in the frequency of application and complexity of chemical control programs. Interdisciplinary solutions to pest problems have only been practical in chemical usage in humid areas where elaborate spray programs have been developed for the simultaneous treatment of diseases, insects, mites, growth regulation, and nutritional deficiencies. In arid areas a less complex spray program may enhance the utilization of cultural practices and biological agents in pest management.

The following examples from New York and Washington point out some of the potential for multidisciplinary studies in humid and semiarid areas.

Pest Management in Humid Areas

The first example deals primarily with apple pests in New York, where the pest situation is one of the most complex found on deciduous tree fruits (Chapman and Lienk, 1971). New York apple growers now use 8 or more chemical sprays to control about 20 important insects and mites and eight to ten sprays for several diseases (Table I). Other chemicals are applied as fertilizers, fruit thinners, growth regulators, rodenticides, and herbicides. Even by combining some of these in single sprays, a grower often makes 15 applications of chemicals to an orchard in a season. This practice is costly, poses a worrisome management problem for the grower, and may inflict undesirable ecological consequences.

Apple researchers and growers in the northeastern United States have taken up the challenge of developing practical pest management systems that minimize ecological damage by chemicals. The approach is to study each major pest segment separately and then to combine these in integrated programs designed to fit each particular ecosystem. To date, much knowledge and technology has been developed for certain specific pests, and other pests are now coming under more intensive study.

Nonchemical Pest Control Methods

The most obvious nonchemical approach to pest control on any crop is to use resistant cultivars. At the present, these are practically unavailable for apple. Cultivars vary in susceptibility to specific pests, but the level of resistance or tolerance is generally insufficient to allow elimination or even relax-

TABLE I. Typical Pesticide Program on Apples in New York State in 1973.

	Winter	April				May				June		July		August		Harvest	Total
		1	2	3	4	1	2	3	4	1 2	3 4	1 2	3 4	1 2	3		
Purpose	Do [a]	Do	ST	GT	HIG	TC	P	BL	PF	1C	2C	3C	4C	5C	6C	pre/post	treatments
Scab				F [b]	F	F	F	F	F	F	F	F	F	F			11
Powdery mildew						M	M	M	M	M	M	M	M				8
Rusts							D	D	D	D	D	D					6
Other fungi				F	F	F	F	F	F	F	F	F	F	F			11
Fire blight								B,B									2
Mites					O----	----O [c]	A,I				A----	----	----	----A			2–4
Aphids					I----	----	----I				I----	----	----	----	----I		2–3
Maggot												I	I		I		3
Plum curculio									I	I							2
Sawfly									I								1
Tarnished plant bug							I		I								1–2
Scales									I								1
W.A. Leafhopper									I	I----	----I		I----	----	----I		1–3
R.B.Leafroller									I	I		I	I				4
Codling moth, etc.										I		I	I				3
Other lepidoptera										I			I				2
Thinning									T	T,I							1
Harvest control													H			H	1–2
Weeds		W					W										1–2
Rodents																R	1

[a] Stages of development: Do—dormant; ST—silver tip; GT—green tip; HIG—half-inch green; TC—tight cluster; P—pink; BL—bloom; PF—petal fall; C—cover sprays

[b] F—fungicide: captan, dithiocarbamates, dodine; M—mildewicide: sulfur, dinocap, morestan; D—dithiocarbamates (also effective against scab); B—antibiotic: Streptomycin; O—oil: superior, 60- or 70-sec viscosity; A—acaricide: many available; I—insecticide: phosphate, carbamate; T—thinning chemical: NAA, NAD, carbaryl; W—herbicides: various; H—harvest control chemicals: Alar, NAA, 2,4,5-TP; R—rodenticide zinc phosphide.

[c] Broken line: one or two treatments during period indicated.

ation of other control measures. This differential sensitivity of cultivars to pests often complicates control within an orchard rather than simplifying it because uniform practices cannot be applied across cultivars. Unfortunately, mixed planting of cultivars in apple orchards is mandated by pollination considerations. However, the benefits of long-term breeding projects are potentially important because commercially acceptable apple varieties resistant to apple scab, maggot, and other pests would permit a more flexible approach to pest management than is now possible. New York and several other areas have had breeding programs for apple scab resistance for decades and resistance has been found, but these programs have not yet yielded cultivars of wide commercial acceptance. However, these initial successes and the current intensity of this research offers promise of scab-free harvests with minimal fungicide usage within the next two decades. This optimism must be tempered by the possibility that new races of the scab fungus could arise to attack the new resistant cultivars, but current breeding programs are designed to minimize such occurrences. There has been less effort on development of resistance to most other plant disease and insect pests of apple, but the tempo of research in this area has recently been increased. The use of horizontal or general resistance to pests in apples should be given greater emphasis in pest management. Significant success may even be achieved by preventing the further introduction of new cultivars highly susceptible to pests through current trends for greater cooperation between apple plant breeders and crop protection scientists.

The most useful of the biological pest control methods used to date in New York State have been the insect pheromones. Pioneering work on these chemicals conducted by the New York State Agricultural Experiment Station at Geneva (Roelofs and Arn, 1968) led to the isolation, chemical identification, and synthesis of pheromones for six important lepidopterous insects of apple. These sex attractants have been used in mass trapping and monitoring of insects. Recent studies indicate that certain insects can be controlled by mass trapping, but this may not be practical under current management programs. However, the use of pheromones to monitor levels of insect populations as a guide for timing and dosage regulation of insecticide sprays is an extremely useful technique for efficient use of chemicals in pest control (Trammel, 1972).

Unfortunately, sex attractants for several important insect species attacking apple, such as the plum curculio and apple maggot, are not yet available and may not exist. Artificial lures have been developed for apple maggot, but these are not as reliable as pheromones because of their relative low order of specificity.

The use of techniques to reduce mite populations on apple in New York by the encouragement of predators has been somewhat limited by the need for intensive insecticide programs to control the lepidopterous insects, apple maggot, and other insect pests. Thus is discouraged the development of desirable microfauna (Trammel, 1972). However, the development of insecticide-resistant parasites and predators may permit chemical control of insects and biological control of mites even in humid areas (Croft and McGroarty, 1973). The opportunity for mite and insect suppression by parasitism and predation is enhanced with the introduction of each new control method for insects that either reduces or eliminates the use of chemicals.

More Efficient Use of Pesticides

A thorough understanding of pest biology (Chapman and Lienk, 1971; Szkolnik, 1969) and epidemiology as well as a knowledge of the practical mode of action of pesticides is essential to the most efficient use of a pesticide. Hamilton *et al.* (1964) worked out the practical mode of action of several fungicides against the apple scab fungus. These have included the protective, after infection, eradicant, and antisporulant properties of fungicides. This information has allowed for efficient use of these chemicals in relation to scab infection periods. Simultaneous studies on epidemiology of several apple diseases now underway in several states should allow more efficient use of fungicides based on disease forecasting and mode of action. Insect monitoring by pheromones should lead to similar efficiencies in insecticide use (Trammel, 1972).

New Pesticide Uses

The use of chemical sprays to manage pests on apples and other deciduous tree fruits may increase in the future even while actual environmental pollution declines. The use of environmentally bland growth regulators to manipulate susceptibility to such problems as fire blight and aphids is a possibility. Other chemicals may be used to reduce the overwintering of pests, inoculum production, and fecundity. For example, several recently discovered compounds reduce sporulation of both the sexual and asexual stages of *Venturia inaequalis*. Thus they decrease overwintering and summer inoculum of scab (Gilpatrick, 1971). The fungicide Difolatan at a high dosage has given extended scab control, which allows a reduction in number of applications in the bloom period from about six to one (Gilpatrick *et al.*, 1971). Certain broad-spectrum pesticides have also been helpful in reducing the total number of pesticides necessary. Dinocap has given simultaneous control of mites and powdery mildew. Benomyl mixed with oil has been extremely effective against several diseases, mites, and a few insects.

The search for new pesticides which more efficiently control pests or offer unique approaches to pest control would seem to be a promising avenue toward improved apple pest management. Chemicals which act as insect repellants and antifeeding agents modify the susceptibility of hosts, have highly selective action, and act systemically within the host would be of unlimited value in pest management and should be searched for diligently.

Strategies for Pest Management in New York on Apples

Management of apple pests in New York must revolve around apple scab until after bloom (Szkolnik, 1969) and apple maggot in middle to late summer (Trammel, 1972). The pink and petal fall periods are also critical (Table I). A strategy developed and in limited use in New York utilizes the single application technique (SAT) with Difolatan at green tip stage of blossom development to control scab until about petal fall in a typical phenological season (Table II). Plum curculio and several lepidopterous insects are controlled by a petal fall spray. Two subsequent insecticide sprays are necessary for apple maggot about six and nine weeks after petal fall. The need for additional sprays is determined by pheromone and visual monitoring observations. Powdery mildew, scab, and other fungi are controlled by fungicides applied in conjunction with the insecticide sprays. This program reduces the usual twelve or more seasonal sprays to five.

A second strategy is proposed if powdery mildew, rusts, rosy apple aphids, or mites are critical problems because of variety, area, or weather (Table III). In this case, a modified single application technique would be used for apple scab. The Difolatan dosage would be reduced by about 40% from the

TABLE II. Reduced Spray Programs for Varieties Not Highly Susceptible to Rust, Powdery Mildew, and Rosy Aphid

Pest	FT	PF	PF + 21 [a]	PF + 42	PF + 63	Number of treatments
Scab and misc. fungi	SAT [b]	F	F	F	F	5
Powdery mildew		M	M	M	M	4
Mites			A--------	--A-------	-A	1–2
Maggot [c]				I	I	2
Plum curculio		I				1
Lepidoptera		I	I	I		3
Misc. insects		I	I			2

[a] PF + 21—21 days after petal fall.
[b] SAT—single application: Difolatan.
[c] One additional maggot spray may be required.
For other footnotes, see Table I.

TABLE III. Reduced Spray Program for Varieties Highly Susceptible to Rust and/or Powdery Mildew and Where Rosy Aphid Control Is Critical

Pest	GT	P	B	PF	PF + 10 [a] days	PF + 21 days	PF + 42 days	PF + 63 days	Number of treatments
Scab and misc. fungi	RSAT [b]	D	D	D	D	D	F	F	8
Powdery mildew		M	M	M	M	M	M	M	7
Rust		D	D	D	D	D			5
Mites-aphids		A,I			A,I------	------	------	------A,I	1–3
Maggot							I	I	2
Plum curculio				I					1
Lepidoptera				I		I			2
Misc. insects				I					1

[a] PF + 10—10 days after petal fall.
[b] RSAT—reduced rate, single application; Difolatan.
For other footnotes, see Table I.

usual SAT rate and followed by a scab fungicide spray at the pink stage of blossom development combined with pesticides for mildew, rust, mites, and aphids. In areas where rust occurs an additional spray at bloom would be necessary. This program then follows the previous program from petal fall onward and would require about eight applications of pesticides per season. Should either of these programs lead to excessive infections of scab on leaves by harvest, a postharvest spray of benomyl is suggested to reduce the overwintering scab inoculum.

The above programs will, no doubt, be more reliable by the development of forecasting methods for pests. Monitoring stations measuring inoculum levels, wetting periods, and infection periods will allow timely application of fungicides for the various diseases. The use of computers to provide rapid determinations is requisite to this approach. Similar computerized monitoring services for insects and mites would put apple pest management on a spray-on-demand basis (Table IV).

Studies have shown that other refinements are possible, which may allow the further reduction in pesticide use such as spraying alternate middles (Trammel, 1972) and eradicating wild and abandoned apple trees or alternate hosts adjacent to orchards. The ultimate will be achieved when chemical pesticides, parasites, predators, host resistance, cultural practices, and other management practices are integrated into a production system that will provide effective pest management.

Obviously, in such a complex management system success will come only

TABLE IV. Monitors for Apple Pests

Pheromone (sex attractants)
Codling moth (*Laspeyresia pomonella*)
Oriental fruit moth (*Grapholitha molesta*)
Lesser appleworm (*Grapholitha prunivora*)
Red banded leafroller (*Argyrotaenia velutinana*)
Oblique-banded leafroller (*Choristoneura rosaceana*)
Fruit tree leafroller (*Archips argyrospilus*)
Bait
Apple maggot (*Rhagoletis pomonella*)
Visual
Mites (various spp.)
White apple leafhopper
Scales (*Lepidosaphes ulmi* and *Aspidiotus perniciosus*)
Aphids (*Aphis pomi*)
Diseases
Weeds
Spore trapping
Apple scab (*Venturia inaequalis*)
Inoculum development
Apple scab (*Venturia inaequalis*)
Apple rusts (*Gymnosporangium* spp.)
Powdery mildew (*Podosphaera leucotricha*)
Weather
Diseases
Insects
None
Plum curculio (*Conotrachelus nenuphar*)
Tarnished plant bug (*Lygus lineolaris*)
Rosy apple aphid (*Anuraphis roseus*)

slowly and piecemeal. Chemicals will be the core of the program for a considerable time, especially in humid areas. But every step completed brings the apple pest manager closer to the reality of growing apples in an environment which includes only a minimum use of pesticides with no serious environmental effects. This will not be fully achieved until we have more resistant varieties, better pesticides, and nonchemical control for certain critical pests. Reduced pesticide programs will undoubtedly allow certain little known pests to achieve major pest status and thus plague this concept (Glass and Lienk, 1971). Tolerance (Glass and Fiori, 1955) and government regulations may eliminate some of our most valuable pesticides. Continuing research and patience will be necessary to overcome all obstacles in the path to practical pest management in apples in humid areas such as New York.

Pest Management in Semiarid Areas

The second example deals with the multidisciplinary aspects of management of mite populations on apple. Much of the information applies to the semiarid growing conditions of eastern Washington, but examples from other areas will also be used.

Mites are generally induced pests on deciduous fruits. Trees are maintained in a high state of vigor during much of the growing season, which creates an ideal substrate for the development of high populations, and the use of broad-spectrum insecticides for insect control frequently eliminates the natural enemies that regulate mite populations at low densities. High populations and frequent applications of acaricides have led to rapid selection of strains resistant to a wide variety of chemicals. In many areas the management of mite populations by the integration of several tactics is essential to avoid serious problems. The following information is necessary to achieve the optimum in management of mite pests of apple.

1. A survey of the phytophagous mite species involved and the principal mortality factors regulating their populations. This study has been conducted for Washington orchards (Hoyt, 1969*a*). *Tetranychus mcdanieli, Panonychus ulmi,* and *Aculus schlechtendali* are the most common phytophagous species on apple, and the primary regulating factor in most orchards is predation by the phytoseiid mite, *Metaseiulus occidentalis*.

2. A means of sampling and estimating populations of prey and predators. A method of counting mite numbers on samples which is adequate for both prey and predators has long been in use (Henderson and McBurnie, 1943). Sampling techniques were developed that provide a practical indication of the population over an area of several acres.

3. Detailed studies of the dynamic interactions of prey and predators. A great deal of research on population development, individual development, distribution, and predation has been completed (Hoyt, 1969*a, b,* 1970; Tanigoshi and Hoyt, unpublished data) and additional investigations are currently under way. This work is being done by entomologists, but systems scientists are cooperating in determining the areas of research.

4. A method of forecasting peak mite populations based on early samples. A simulation model of mite population development is currently being prepared by systems scientists utilizing the information on population dynamics.

5. A means of determining the damage that the predicted mite population may cause through the establishment of economic injury levels. Chapman *et al.* (1952) and Lienk *et al.* (1956) conducted research on the effects of mite feeding damage on apples, but their studies did not provide quantitative information on the effects of various population levels. Research is currently being con-

ducted on the effects of various levels of mite populations on several fruit quality factors, fruit color, size, and total yield. These studies involve a cooperative effort between plant physiologists and entomologists.

6. The economic consequences of the potential loss to the grower, the fruit industry, and the consumer. This would require the ability to predict yield and price, and would involve a cooperative effort by economists and systems scientists. Numbers 4, 5, and 6 would be used together in determining cost-benefit ratios and the tactics to be employed in regulating populations.

7. Alternative tactics for managing mite populations. The two alternatives of biological control and selective acaricides are the primary tactics presently used for mite control (Hoyt, 1969*b*). Other possible methods of reducing mite populations may be developed through multidisciplinary studies as described below.

8. The effect of cultural practices on the development of mite populations, the damage caused by the mites, and on the mite–predator interaction. It is in this area where the greatest opportunity for multidisciplinary studies occurs. In the following paragraphs the effects of various cultural practices on mite populations and damage are discussed.

The selection of apple varieties is dictated largely by the demands of the market. Although long-term breeding programs are under way to develop insect- and disease-resistant varieties, there is presently little indication that resistance to pests would be given much consideration in the selection of commercial varieties on a large scale. This appears to be particularly true with mites. The greatest potential for growth of populations of *Tetranychus urticae* (Bengston, 1970) and of *Panonychus ulmi* (Ghate and Howitt, 1965) is on Delicious, the most popular variety of apple. Downing and Moilliet (1967) also found leaves of the Delicious variety to harbor fewer predacious mites than several other varieties.

The design of an orchard is determined by labor requirements, varieties to be planted, irrigation practices, location, yield, and currently accepted practices. The design of an orchard in relation to mite population management has not been studied. A shift from older plantings of large trees (about 50 trees per acre) to small, densely planted trees (from 250 to 500 per acre) could have a substantial effect on mite numbers, damage, or predator–prey interactions. These effects could result from reduced overwintering sites on small, smooth-barked trees, greater exposure of predators to hazards from sprays, or changed economic injury levels.

Irrigation practices in orchards are currently undergoing major changes. Overhead irrigation systems have become popular in western apple-growing areas because of reduced needs for labor and the potential for frost and climate control. Overhead irrigation brings on microclimatic changes which may affect pests, or pesticide deposits may be reduced by the sprinkling. Hudson and

Beirne (1970) have shown that overtree sprinkling is effective in maintaining populations of *T. mcdanieli* at low levels, but it had less effect on *P. ulmi* populations. This could result in a change in the relative importance of the two species. Trickle irrigation is also being tested in several orchard areas. This method of irrigation may have a substantial effect on cover crops and indirectly on pest and beneficial insect populations because only the area immediately adjacent to the tree is irrigated.

Inadequate irrigation may produce different effects depending on the species of mite involved. Specht (1965) showed that water stress significantly reduced the population growth of *P. ulmi,* while several authors have reported that drought conditions may make plants more suitable for reproduction by species of the genus *Tetranychus*. Lack of water may also cause cover crop hosts of mites to dry and force the mites to move into the trees. Trees which are under water stress also cannot tolerate as high a mite population as those receiving adequate irrigation.

So far, little or no attempt has been made to include the management of mites as a part of the selection of the optimum irrigation method. It would probably be unwise to recommend overhead irrigation as a favorable practice in mite population management until more is known about its effects on the total pest complex.

Pruning methods are designed primarily to maintain maximum yield of high quality fruit and to control tree size and shape. Pruning does remove some overwintering stages of *P. ulmi,* certain predators, and *Aculus schlechtendali,* an alternate source of food for predators, but the effect on mite populations appears to be temporary. In Pennsylvania, it is recommended that suckers be left in trees until June to conserve eggs of several predators which are deposited in this tissue (Asquith, 1969). Proper pruning and tree training may have secondary effects on mites or mite damage. Heinicke (1966) has shown that color development in fruit is directly related to exposure to sunlight. Good fruit exposure may therefore offset some mite feeding damage. Proper pruning will also facilitate better spray coverage, which may be beneficial to the total pest control program.

Several studies on the effects of plant nutrients on mite populations have been conducted. These are reviewed by van de Vrie *et al*. (1972). In most studies a direct relationship existed between levels of leaf nitrogen and mite populations. In addition, well-fertilized trees tended to have a prolonged vegetative period allowing mites to build up to higher densities (van de Vrie, 1973). However, trees that are properly fertilized can probably tolerate greater mite numbers. Higher populations with a higher host tolerance are probably advantageous to predation since predators are less likely to overexploit their prey. Excessive nitrogen levels can lead to poor color development and decreased storage life of fruit (Weeks *et al*., 1952), so moderate, early applications of nitrogen appear to be optimum for crop management.

The relationship of levels of other nutrients to mite populations is not well understood at present. There are conflicting reports on the effects of phosphorus and potassium on mite populations, but the use of metal chelates limits population increases of *Tetranychus urticae* on beans (Cannon, 1970).

Cover crop management and proper weed control are important in the regulation of mite populations. Where good sod cover crops are maintained in the Northwest, mite populations generally remain at lower levels than where clean cultivation is practiced (Hoyt and Caltagirone, 1971). This may be due in part to the control of dust which favors mites and inhibits predation. In New Jersey integrated control has been most successful where sod cover was maintained in the middles but where herbicides removed all grass under the tree, which left scattered broad-leaved weeds such as dock and dandelion (Christ, 1971). In areas where *Stethorus punctum* is an important mite predator, cultivation under trees should be avoided until mid-June to give hibernating beetles a chance to emerge before the soil is disturbed (Asquith, 1971).

Herbicides may have little direct toxicity to mite predators, but the destruction of weeds may eliminate populations of mites and predators outside the tree. This may be relatively unimportant where most of the predator–prey interaction takes place in the tree (Hoyt, 1969*a*), but in other areas could destroy an important source of predators which would later move into the tree (Croft and McGroarty, 1973). These authors point out the potential of herbicide treatments, host plant types, and mowing practices for manipulating prey and predator populations in ground cover. Elimination of all weeds may be undesirable because it simplifies the ecosystem and creates instability. However, much work needs to be done to determine which species of cover crop and which management techniques are most favorable to management of the pest complex.

Fruit thinning is generally accomplished by the use of chemical thinners followed by some hand removal of fruit. Most chemical thinners have little effect on mite populations, but carbaryl is highly toxic to predators and its use can result in a rapid increase in mite numbers. Since the use of carbaryl as a thinner and integrated mite control were both considered essential, a study involving plant physiologists and entomologists was undertaken to develop a satisfactory method of use. A selective spray technique was developed which gives the desired thinning and conserves predacious mites (Hoyt, 1969*b*).

Spraying is one cultural practice where multidisciplinary studies are essential to sound pest management. This is because of the devastating and overriding effects a spray may have, which may negate all other management tactics. Sprays can influence mite populations in several ways. They can directly destroy populations of important predators, leaving mites to reproduce unchecked by natural controls. They can excessively reduce populations of prey or alternate sources of food for predators resulting in a later rebound of destructive populations, or they can produce a better food supply for mites and prolong its availability by controlling pests or altering the physiology of the tree. For these reasons it is important to determine the total effects of each agricultural chemical. For example, in the eastern United States it would be impossible to study the management of mite populations without relating this to the extensive disease control programs. Even in the West, where disease problems are less severe, sprays for the control of apple powdery mildew may interfere with the regulation of mite populations.

Wherever possible, the use of nonchemical tactics for the control of insects and disease should prove beneficial to the management of mite populations. Where chemicals are necessary it is important to choose the chemical, dosage, timing, and application technique which provide the greatest degree of selectivity to achieve the optimum in long-term regulation of pest populations.

Implementation of all of these tactics which suppress mites or enhance the activity of predators would lead to greatly reduced needs for acaricides. This would probably require extensive education of growers to point out the potential of these cultural practices in the management of pests.

The discussion of this section has considered only the management of mites. If this were expanded to include the many other disease, insect, and weed pests that attack apples, the complexity of the system and the multidisciplinary efforts required to devise sound management systems would increase greatly. In essence an analysis of the orchard ecosystem would be required.

Literature Cited

Asquith, D., 1971, The Pennsylvania integrated control program for apple pests—1971, *Penn. Fruit News* **50:**43–47.

Asquith, D., and Horsburgh, R. L., 1969, Integrated versus chemical control of orchard mites, *Penn. Fruit News* **48:**38–44.

Bengston, M., 1970, Effect of different varieties of the apple host on the development of *Tetranychus urticae* (Koch), *Queensl. J. Agr. Anim. Sci.* **27:**95–114.

Cannon, W. N., Jr., 1970, Population dynamics of *Tetranychus urticae* on bean leaves treated with metal chelates, *J. Econ. Entomol.* **63:**722–725.

Chapman, P. J., and Lienk, S. E., 1971, Tortricid fauna of apple in New York, Special Publication, New York State Agricultural Experiment Station. 122 pp.

Chapman, P. J., Lienk, S. E., and Curtis, O. F., 1952, Responses of apple trees to mite infestations, *I. J. Econ. Entomol.* **45:**815–821.

Christ, E. G., 1971, 1971 Tree fruit production recommendations for New Jersey, New Jersey Extension Service Leaflet 446-A. 50 pp.

Croft, B. A., and McGroarty, D. L., 1973. A model study of acaricide resistance, spider mite outbreaks, and biological control patterns in Michigan apple orchards, *Environ. Entomol.* **2:**633–638.

Downing, R. S., and Moilliet, T. K., 1967, Relative densities of predacious and phytophagous mites on three varieties of apple trees, *Can. Entomol.* **99:**738–741.

Ghate, A. V., and Howitt, A. J., 1965, Mite resistance to organophosphorous compounds and the response of apple varieties to mites in Michigan, *Quart. Bull. Mich. State Univ. Agr. Exp. Sta.* **47:**322–350.

Gilpatrick, J. D., 1971, New developments in the control of apple scab, *Proc. Mass. Fruit Growers Assoc.* **77:**25–32.

Gilpatrick, J. D., and Szkolnik, M., and Gibbs, S. D., 1971, A single high rate application of Difolatan for the control of apple scab, *Phytopathology* **61:**893.

Glass, E. H., and Fiori, B., 1955, Codling moth resistance to DDT in New York, *J. Econ. Entomol.* **48:**598–599.

Glass, E. H., and Lienk, S. E., 1971, Apple insect and mite populations developing after discontinuance of insecticides: 10-Year record, *J. Econ. Entomol.* **64:**23–26.

Hamilton, J. M., Szkolnik, M., and Nevill, J. R., 1964, Greenhouse evaluation of fruit fungicides in 1963, *Plant Dis. Reptr.* **48:**295–299.

Heinicke, D. R., 1966, Characteristics of McIntosh and Red Delicious apples as influenced by exposure to sunlight during the growing season, *Proc. Am. Soc. Hort. Sci.* **89:**10–13.

Henderson, C. F., and McBurnie, H. V., 1943, Sampling techniques for determining populations of the citrus red mite and its predators, *U.S. Dept. Agr. Circ.* **671:**1–11.

Hoyt, S. C., 1969*a*, Population studies of five mite species on apple in Washington *in:* Proceedings of the Second International Congress on Acarology, Sutton-Bonnington, England, 1967, pp. 117–133.

Hoyt, S. C., 1969*b*, Integrated chemical control of insects and biological control of mites on apple in Washington, *J. Econ. Entomol.* **62:**74–86.

Hoyt, S. C., 1970, Effect of short feeding periods by *Metaseiulus occidentalis* on fecundity and mortality of *Tetranychus mcdanieli, Ann. Entomol. Soc. Am.* **63:**1382–1384.

Hoyt, S. C., and Caltagirone, L. E., 1971, The developing programs of integrated control of pests of apples in Washington and peaches in California, *in:* Biological Control (C. B. Huffaker, ed.) New York, Plenum Press, pp. 395–421.

Hudson, W. B., and Beirne, B. P., 1970, Effects of sprinkler irrigation on McDaniel and European red mites in apple orchards, *J. Entomol. Soc. Brit. Col.* **67:**8–13.

Lienk, S. E., Chapman, P. J., and Curtis, O. F., 1956, Responses of apple trees to mite infestations. II., *J. Econ. Entomol.* **49:**350–353.

Roelofs, W. L., and Arn, H., 1968, Sex attractant of the red-banded leaf roller moth, *Nature* **219:**513.

Specht, H. B., 1965, Effect of water-stress on the reproduction of European red mite, *Panonychus ulmi* (Koch) on young apple trees, *Can. Entomol.* **97:**82–85.

Szkolnik, M., 1969, Maturation and discharge of ascospores of *Venturia inaequalis, Plant Dis. Reptr.* **53:**534–537.

Trammel, K., 1972, The integrated approach to apple pest management and what we are doing in New York, *Proc. N.Y. State Hort. Soc.* **117:**37–49.

van de Vrie, M., 1973, De fruitspintmijt en andere mijten op vruchtbomen, Publ. Research Station for Fruit Growing, Wilhelminadorp, No. 13, 67 pp.

van de Vrie, M., McMurtry, J. A., and Huffaker, C. B., 1972, Ecology of tetranychid mites and their natural enemies: A review. III. Biology, ecology, and pest status, and host-plant relations of tetranychids, *Hilgardia* **41:**343–432.

Weeks, W. D., Southwick, F. W., Drake, M., and Steckel, J. E., 1952, The effect of rates and sources of nitrogen, phosphorus, and potassium on the mineral composition of McIntosh foliage and fruit color, *J. Am. Soc. Hort. Sci.* **60:**11–21.

X

Integrated Forest Pest Management: A Silvicultural Necessity

William E. Waters and Ellis B. Cowling

Introduction

Diseases and insects are major causes of loss in the value and productivity of forest resources. Prevention or reduction of these losses by adjustments in forest management and utilization practices is a familiar concept in forestry; but the concept has yet to be implemented fully.

Most forest resource managers have viewed disease and insect losses as more or less random and unpredictable events. As a consequence, there has been a strong tendency to "wait-and-see" and to make short-term decisions about how to solve immediate pest problems after they develop, rather than to anticipate them and to take preventive action before the losses occur. Some pests that cause obvious damage, mortality, or conspicuous losses in growth have been long recognized. But many pests that cause inconspicuous damage have been ignored even though the cumulative growth impact of such pests can be substantial.

When serious insect-caused losses have been detected, forest entomologists generally have responded to the concern of forest managers by

---

WILLIAM E. WATERS · Pacific Southwest Forest and Range Experiment Station, U.S. Forest Service, Berkeley, California. Present address: College of Natural Resources, University of California, Berkeley, California 94720. ELLIS B. COWLING · Departments of Plant Pathology and Forest Resources, North Carolina State University, Raleigh, North Carolina 27607.

recommending salvage cuttings and chemical-spray programs aimed at temporary suppression of the offending population of insects. When disease-caused losses have occurred, forest pathologists generally have responded with management recommendations that would prevent losses in future rotations but often would do little for the existing stand. Neither response has been entirely satisfactory, and communication between forest managers and pest research and advisory personnel continues to be irregular and often ineffective.

During the past three decades, demands for forest products and nontimber uses of forest lands have increased enormously. This has provided impetus for increased communication and cooperation among forest entomologists and pathologists, other specialists in the sciences basic to forestry, practicing forest managers, and pest control personnel. The result has been increased efforts to develop better methods for detection and evaluation of losses due to forest insects and diseases as well as to develop greater understanding of the population dynamics of destructive insects, the epidemiology of disease, and the genetics of pest susceptibility and resistance. Only occasionally, however, have reliable economic analyses been made of disease- and insect-caused damage or of alternative management recommendations.

Today, taxpayers are demanding that public investments in forestry research and extension programs pay public dividends by increasing timber production, by increasing water and wildlife yields, and by making forest lands more attractive and suitable for recreational purposes. Thus, it is timely, even urgent, that we systematically integrate developing knowledge about forest insects and diseases into reliable forest resource management practices.

What is needed are flexible resource management systems that are ecologically and economically sound; systems that will permit optimal value and utility to be derived from forest resources at minimum cost; systems that will meet future as well as current management needs irrespective of the type, age, and present quality of the forest; systems that will assure minimum losses due to forest pests with minimum adverse effects on the quality of the environment and the long-term productivity of forest ecosystems.

The purposes of this paper are to describe briefly: (1) the distinctive characteristics of forest ecosystems and the nature of forests as renewable resources for man; (2) the components of a forest pest management system and how they interrelate in the processes of management planning and decision making; (3) suggested steps for the development of effective and efficient pest management systems; (4) the status and practical limitations of available management plans for diseases and insects in three major types of forests in the United States; and finally (5) to offer a series of recommendations for the improved integration of the pest management concept with the practice of forest resource management.

The Forest as a Dynamic Ecosystem

One-third of the 1.1 billion hectares (2.3 billion acres) in the United States is forested (U.S. Forest Service, 1973). Ecologically, this ranges from mesquite brushland, mangrove swamp, and spruce bog (where the trees are barely identifiable as such) to the magnificent coniferous forests of the Pacific Coast and the colorful hardwood stands of New England. One gets some measure of the diversity of forest ecosystems by viewing the changes that occur with elevation in the major mountain ranges of the western and eastern United States. Here, one can pass from subtropical desert to alpine meadow, or from bottomland swamp to heathlike tundra, in a single day's climb. The mix of plant and animal species one observes may be comparable to a longitudinal spectrum of 1000 kilometers or more.

The forested areas of the United States are classified according to various botanical, edaphic, climatic, and socioeconomic criteria (U.S. Department of Interior, 1970; Society of American Foresters, 1964). Most descriptions are essentially static and do not adequately indicate the true diversity, complexity, and changing nature of forests as ecosystems. Dynamic forces are at work in every hectare and it is the continuing interplay of the biosynthetic and destructive forces—the cumulative *net* effects of these competing forces—that determines the character and productivity of a forest over time.

Insects and diseases are the most destructive agents affecting forests in the United States (U.S. Forest Service, 1958, 1973). They attack all parts of living trees in all of their life stages from seed to harvest (or natural demise). Their depredations—alone and in combination with fire, drought, wind, and other destructive agents, including man—directly influence the composition, structure, growth, and regeneration of forest stands. From an ecological standpoint, they exert both limiting and disruptive effects on the ecosystems in which they operate.

Some diseases and insects can cause irrevocable—or at least very long-term—changes in forest ecosystems; for example, they can virtually eliminate an entire tree species. Chestnut blight, Dutch elm disease, and red-pine scale have had this effect in many areas where they have become established. Other insects and diseases (e.g., balsam woolly aphid, white pine blister rust, dwarf mistletoes, and fusiform rust of southern pines) usually deform or kill many, but not all, of their host trees in a given area. Such pests can greatly modify the composition and age–size structure of affected stands. Epidemic insects such as the Douglas-fir tussock moth, spruce budworm, forest tent caterpillar, and southern pine beetle can cause much damage in a short period of time, but their effects are not irreversible. Attacked stands often regenerate the same tree species, and only the time sequence of stand development is altered.

Less spectacular pests, rarely abundant enough to be considered serious by themselves, may cause important cumulative effects on forest ecosystems. For example, the complex of oak leaf rollers and associated defoliating insects in eastern oak forests—particularly when coupled with root diseases—continuously deteriorate host stands over extensive areas. Many twig and bud borers attacking young conifers and hardwoods cause stunting, multiple stems, and similar defects so regularly that the typical form of their host trees is modified greatly.

Our knowledge of the ecological interplay among pest incidence, pest control, and silvicultural practice is still incomplete. We know from research and experience that some silvicultural practices increase either the abundance of pests or the probability of damaging attacks by them. Similarly we know that the most effective silvicultural method to prevent or decrease the amount of damage by one destructive agent may be ineffective, or even stimulatory, for another agent affecting the same tree species. For example, fire suppression programs have led to greatly increased abundance of the alternate (oak) hosts for fusiform rust and have also increased the prevalence of the brown-spot disease of longleaf pine. A given silvicultural treatment designed to promote rapid growth or other desirable attributes in a given forest stand may also increase the hazard of loss caused by disease or insects. Examples of this include the increased prevalence of annosus root rot after thinning and of fusiform rust after fertilization of southern pine forests. These anomalies must be resolved by better understanding of how trees grow, how pathogens and insects interact, and how various silvicultural treatments affect the complex interactions of trees with their insect and disease parasites (Waters, 1969, 1972). In many cases, knowledge of natural ecosystems will provide useful guides for pest control in managed forests (Dinus, 1974).

The Forest as a Renewable Resource for Humans

Forests provide a renewable resource for many needs of our society: lumber, plywood, and other forest products for homes; wooden furniture, implements, and a myriad of other items of everyday use; paper products in prodigious amounts—more than 600 pounds annually for every man, woman, and child in the United States; cellulose and other wood-based chemicals including such specialty products as turpentine and maple sugar.

Forests also provide food and habitat for birds, fish, and many game animals. They provide protective cover and regulate stream flow in our watersheds. They also help replenish oxygen in the atmosphere and provide metabolic sinks for the removal or modification of atmospheric pollutants. Not the least of their benefits, forests provide places of beauty and recreation where we can find rest, inspiration, and a refuge from the tensions and frustrations of life.

Man must learn to exercise a more responsible dominion over these resources. We must harmonize our needs and aspirations with the finite capacity of forests to provide the various benefits desired from them.

In practice, certain social, political, and economic constraints make optimization of forest resource values and benefits difficult to realize:

1. Three major economic facts-of-life have an important bearing on investments in forest pest management: (a) forests provide only small annual increments of value per unit of land area, typically only $20–75 per hectare per year; (b) the total values accumulated over the full time-span of a forest crop can be quite large, often $2000–4000 per hectare, and in some cases as much as $100,000 or more; and (c) where timber values are paramount, many years or even decades may elapse between the time a given silvicultural or pest management treatment is applied and the recovery of that investment at the time of harvest.

2. Forests are owned and managed by many different persons and organizations with a great diversity of management goals. Private individuals and corporations own about 73% of the commercially productive forest land of the United States; the remainder is in federal, state, and other public holdings (U.S. Forest Service, 1973). About 20% of the commercial forest area is in holdings of 25,000 or more hectares and is managed by about 300 large corporations and government agencies. But 30% is in small woodlots of 50 or fewer hectares owned by nearly four million private individuals (Duerr *et al.*, 1974). Obviously, implementation of comprehensive pest management programs is much easier to achieve on the few larger than on the many smaller holdings.

3. Federal and state taxation policies and forest development programs significantly influence the financial attractiveness of certain pest management activities. An array of general property taxes, yield taxes, severance taxes, and income and capital gains taxes are levied against private owners of forests by local, state, and federal governments. The federal income tax, and especially its capital gains provisions, increases incentives for pest management investments by allowing these expenses as a deduction from gross income. Timber severance taxes, on the other hand, are levied directly per unit volume of production and may discourage harvest of diseased and otherwise inferior trees. Progressive taxes like the federal income tax tend to favor pest management investments on larger holdings more than on smaller ones. However, the recently enacted Forestry Incentives Program (P. L. 93-86, Title X) provides direct subsidies of 50–75% of the cost of certain timber stand improvement efforts that may reduce disease and insect losses on small holdings.

4. Most nontimber values of forests are difficult to quantify. Estimated

prices for harvested timber in the current and future rotations can be used in calculating benefits and costs for various pest management practices affecting timber production. It is much more difficult to develop rational benefit-cost estimates for the nontimber values associated with water yields and quality, wildlife populations, and the aesthetic and other amenity values of forest ecosystems.

Social values are also making increasingly important impacts on forest pest management practices just as they are on those for agricultural crops. Pesticide use and clear cutting in forest areas are currently controversial. Social concerns increasingly are being expressed in statutory constraints and legal actions.

Forest Pest Management Systems

Basic Components

The basic principles of forest pest management, and the major components of the decision-making process, are essentially the same in forestry as in agriculture. These principles have been outlined in an earlier chapter in this volume. But the time- and space-frame of forest management—the continuity of forest cover in a given area and the commitment of management to optimize the multiple resource values involved over long periods of time—requires additional information for planning and action. Also, a greater variety of biological conditions and a wider range of economic and social values must be considered in forestry than in agricultural pest management.

Figure 1 shows the general structure of an integrated forest pest management system as adapted from Waters and Ewing (1975). This figure depicts both the research and development phase (enclosed by the dotted lines) and the operational phase (outside the dotted lines) of an effective system for management of pests in a forest ecosystem. The research and development core includes four basic components: (1) the population dynamics and epidemiology of the insects and diseases present, (2) the dynamics of forest stand development, (3) the socioeconomic impacts of pest-caused damage on resource values, and (4) the treatment strategies. The latter two components provide direct inputs for benefit-cost analysis; this is the basic mechanism of decision making in the operational phase of the system. Each component is a complex subsystem in itself. The linkages (or information flows) among the components are shown by the heavier arrows, and the feedbacks are indicated by the lighter arrows. In modeling this system, both sets of arrows indicate the components that provide inputs that are necessary parameters for the other components.

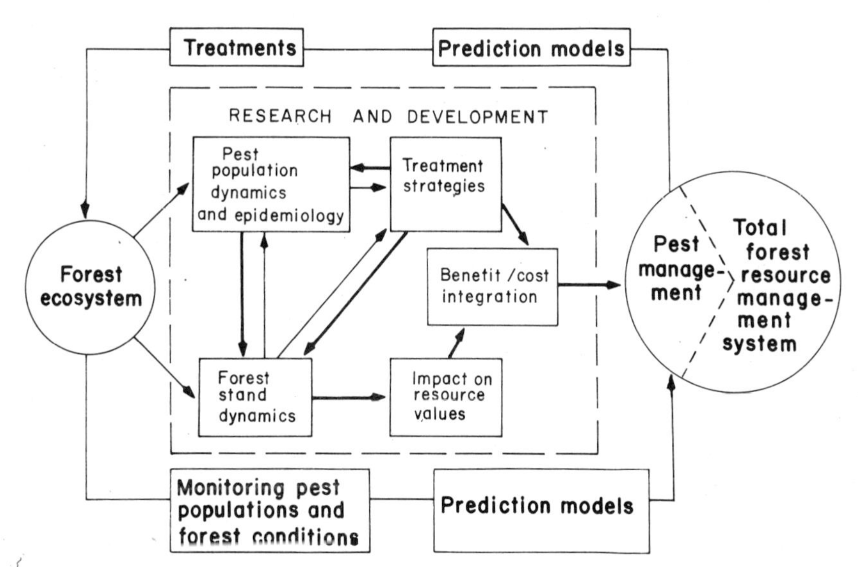

Figure 1. Model structure of a pest management system. The research and development phase is shown inside the dotted lines and the operational phase outside the dotted lines.

Regularly coordinated modeling of both pest populations and forest stand conditions is required in both the operational and research and development phases of the system. Monitoring of insect populations usually is based on direct sampling of their numbers or indirect estimates of relative densities determined from an analysis of damage symptoms. Monitoring of tree diseases usually takes the form of disease hazard evaluations which are based on the incidence of disease-caused damage or on analyses of soil or meteorological factors known to influence the infection potential of the pathogen(s) involved.

In the operational phase of the system, prediction models are the mechanism of projecting pest population and/or stand conditions ahead in time on the assumption of no treatments or intervention by man (bottom line of Figure 1); they also provide the means for evaluating the outcomes of alternative treatment stategies in terms of relative effectiveness and efficiency, i.e., benefit and cost (top line of Figure 1).

Steps in the Development of an Adequate Pest Management System

The major steps necessary in the development of an adequate system of forest pest management may be summarized as follows:

1. Define and quantitatively describe: (a) the forest ecosystem to be protected, (b) its extent and boundaries, (c) the resource values to be derived from it, and (d) the objectives for management of the ecosystem in the short- and the long-run. Identify the major and minor land owners or land-owner groups and their management personnel. Identify and characterize the management practices presently being used or contemplated for specific areas within the ecosystem.

2. Identify the significant destructive agents (insects, diseases, animals, weed pests, fire, weather, etc.) which limit the productivity or negatively influence the value of the resources to be derived from each tree species in the ecosystem. Determine the potential biological and economic impact of each destructive agent.

3. From analyses of the dynamics and epidemiology of the pest(s) identified under (2) above, determine the critical biotic and abiotic factors that affect their distribution and changes in their numbers and the amounts of damage they cause. Determine the potential of these factors for effective control or regulation of each pest acting alone or in combination with others. Develop prediction models of pest numbers and/or damage utilizing measurements or observations that are practically feasible in an operational management system.

4. Obtain or develop prediction models for each tree species or ecological group of host plants in the ecosystem and modify these models as needed to include pest-related parameters and effects. These models must be of a form to incorporate projections of mortality and growth changes due to insects and diseases over time. Develop practical monitoring procedures for an operational survey system.

5. From (4) above, develop quantitative methods for the prediction and evaluation of the potential impacts of the pests on specific resource uses and values.

6. From (3) and (4) above, test and develop a variety of techniques and strategies to prevent, suppress, ameliorate, or regulate pest populations and/or the damage they cause. Develop biological and economic thresholds for each technique and strategy and analytical means of coupling the biological thresholds for applying these treatments (i.e., at what point in the development of a pest infestation they will be most effective) with the economic thresholds of damage for the tree species involved. Develop methods for benefit-cost analysis of these treatments.

7. Identify any legal or social constraints on the use of potential management treatments and practices and incorporate this information in the models for management planning and decision making.

8. Integrate all the foregoing into a cohesive and flexible decision and planning process. Test and validate all procedures and models involved under

operational conditions. Prepare practical guidelines for operational use and provide them to the managers identified under step (1) above.

Current Practices and Future Needs

A number of empirical guidelines have been developed for use in deciding when, where, and how to control specific forest insects and diseases (Marty and Mott, 1964; Marty, 1966; Kuhlman *et al.*, 1975; Campbell and Copeland, 1954). Most of these guidelines are limited in area of application and/or methods of control; often they are based on less than adequate knowledge of the biology and/or the economic impact of the pest concerned. None of these guidelines now encompass all of the components essential for a fully adequate pest management system. At present, several comprehensive research programs aimed at the development of management strategies for specific forest pests* are being developed. These include interdisciplinary teams of systems analysts, biologists, and economists whose job is to put the whole package together into practical form. This is a step in the right direction.

The current status of forest pest management in the United States and future needs for practical development are perhaps best shown by specific examples.

The Old-Growth Douglas-Fir Forest

First, we will look at the major insect and disease problems in an old-growth forest where production of timber is the major goal, the trees are mostly mature or overmature, and protection is aimed mainly at keeping the trees alive and as free of defects as possible until they can be harvested. The old-growth Douglas-fir forests of the Pacific Coast states are a good example.

Douglas-fir is the most important softwood lumber species in the United States. It makes up more than one-fourth of the nation's current total sawtimber volume; 60% of this is in the Pacific Coast states (U.S. Forest Service, 1973). Old-growth stands cover about one-fourth of the 9 million hectares in commercial Douglas-fir timberland in these states; and they contain about one-half of the total growing stock of Douglas-fir.

The pests causing the greatest losses in old-growth Douglas-fir are *Fomes*

* Notably the gypsy moth, western pine beetle, mountain pine beetle, southern pine beetle, Douglas-fir tussock moth, and dwarf mistletoes.

pini and other heartrot fungi, the Douglas-fir beetle (*Dendroctonus pseudotsugae*), and the root rot caused by *Poria weirii*. In addition, two defoliating insects, the western spruce budworm (*Christoneura occidentalis*) and the Douglas-fir tussock moth (*Orgyia pseudotsugata*), recur periodically in epidemic numbers and cause widespread damage to both old-growth and second-growth stands of Douglas-fir.

Fomes pini. This fungus causes the most destructive single disease of old-growth Douglas-fir. It attacks vigorous trees as well as suppressed ones and causes mortality as well as heartrot. The decay is a distinctive white-pocket type which develops in several species of conifers in addition to Douglas-fir. Serious losses also are inflicted by other heartrot fungi; these include *Fomes officinalis, F. subroseus,* and *Polyporus schweinitzii*. These fungi cause a brown cubical type of decay which decreases the strength of the wood much more rapidly than the white-rot fungi such as *Fomes pini*.

The white pockets in the wood attacked by *Fomes pini* contain soft white fibers separated by zones of apparently sound wood. In the incipient stage, strength is not affected significantly; in the advanced (white-pocket) stage, however, the amount of strength loss is roughly proportional to the abundance of the white pockets. This fungus has never been found to cause decay in wood products after harvest.

The amount of heartrot in Douglas-fir increases progressively with increasing age of the trees (Boyce, 1932). As Table I shows, a very substantial shift takes place after 160 years of age. Up to that time biosynthetic forces predominate in the development of Douglas-fir trees, whereas after that time biodeterioration gains the upper hand. The obvious way to minimize loss due to heartrot is to harvest the trees prior to this shift. For reasons of accessibility, current market conditions, etc., this is not always feasible. As a result, alternative procedures for minimizing disease losses must be considered: (1) harvesting the trees before heartrot losses become excessive and (2) utilization of partially decayed wood.

TABLE I. Relation of Heartrot to Age in Douglas-Fir

Tree age (years)	Amount of decay (% of volume)
<40	0
40–80	2
80–120	4
120–160	9
160–200	34
>200	85

The amount of decay in a given stem can be estimated from the approximate age of the tree and the number and position of sporocarps and so-called "punk knots" produced by the decay fungi on the bole of the tree. Thus, in selective harvesting operations, first preference should be given to cutting the oldest trees with the largest numbers of external indicators of internal defect.

During the past several years, the technical utilization requirements for Douglas-fir have been adjusted to permit utilization of logs partially decayed by *Fomes pini*. Plywood containing white-pocket veneers is used as decorative paneling, and in packaging crates, sheathing, and other products where structural integrity and/or attractiveness are desired rather than maximum strength. Use of both sound and partially decayed veneer sheets in the same plywood composite is also permitted; these new specifications of the Douglas Fir Plywood Association have provided a major stimulus for utilization of logs infected with *Fomes pini*. In addition, pulp and paper products made from wood partially decayed by this fungus are satisfactory for many purposes. Reasonable yields of good-quality pulp can be obtained from logs too decayed for utilization as plywood. In fact, paper sheets made from white-pocket logs are stronger in tensile strength, although somewhat weaker in tear resistance, than papers made from nondecayed logs. The heartrot caused by *Fomes pini* well illustrates that marketing and technical utilization considerations as well as biological factors must be considered in developing optimal management strategies to minimize losses due to forest pests.

Douglas-Fir Beetle. This insect normally kills weakened and overmature trees. Outbreaks usually develop in areas where a significant number of trees have been damaged by wind, ice, fire, defoliation, or logging activity, or where much cut material has been left on the ground after logging (Furniss and Orr, 1970). The beetle populations spread from these especially suitable breeding places to adjacent healthy stands.

It has not been feasible to control the beetle with insecticides. Effective removal of damaged trees prior to invasion by the beetles often is limited by accessibility and other logistic considerations. Attempts to log damaged or fallen timber promptly or to log trees while the beetles are still in them, often create stand disturbances and additional injuries that induce further beetle attacks and outbreak conditions. Recent research has indicated that a potent antiattractant pheromone, methylcyclohexenone, may be useful in preventing mass attacks in susceptible timber (Rudinsky, 1968; Furniss *et al.*, 1972). In addition, an attractant pheromone, frontalin, in combination with volatile substances produced by the host, has shown some promise for use in trapping and/or diverting beetles from high hazard areas (Furniss *et al.*, 1972; Knopf and Pitman, 1972). These latter approaches may be the most effective direct means of preventing beetle attack and population buildup in the future, but more field trials are needed to demonstrate their efficacy and practicality. At present, the conversion of the old-growth stands to young vigorous ones, coupled with improved access, appears to provide the most feasible means of decreasing losses due to the Douglas-fir beetle.

Western Spruce Budworm and the Douglas-Fir Tussock Moth. These two major defoliating insects differ greatly from the Douglas-fir beetle in their biology, patterns of occurrence, and damage. But essentially the same economic factors can be used in making decisions about if, when, and where to take action against them.

The budworm is the most widely distributed and destructive defoliator of coniferous trees in western North America (Carolin and Orr, 1972). It is especially damaging to old-growth Douglas-fir. Outbreaks often develop slowly but extend over large areas. The frequency and duration of outbreaks varies considerably.

This insect has a complex feeding habit; it begins with needle- and bud-mining by the young larvae and more general feeding on the foliage in the later larval stages. It feeds primarily on the new foliage of the current year, so it generally takes 2–3 years of heavy feeding to cause a significant decrease in growth and 4–6 years to kill trees. The stress of budworm defoliation on old-growth trees greatly increases their susceptibility to the Douglas-fir beetle, Poria root rot, and other pests.

At present, there are not many options for control of the spruce budworm. Although over 40 species of parasites attack it, even their combined effects are not sufficient to hold it in check when climatic conditions are favorable for a general increase in population densities (Carolin and Orr, 1972). For this reason, and the diverse ecological conditions under which the budworm occurs, neither augmenting the native parasite complex nor introducing new ones has been considered feasible. Similarly, the prospect for prevention or amelioration of outbreaks by silvicultural means is not very hopeful. The budworm attacks trees of all ages, its host species are often intermingled, and most critical, perhaps, adjacent owners with different management objectives make it practically impossible to coordinate this approach over a sufficiently large area to be effective.

Direct suppression of increasing budworm populations can be achieved by use of insecticides. Two chemicals, Malathion and Zectran, presently are registered for this purpose. Recent field tests with commercial formulations of *Bacillus thuringiensis* against the eastern form of the budworm have been promising; thus, this bacterial insecticide may also be added to the arsenal soon. These materials, chemical and microbial, all require careful timing of application and good coverage to be effective, requirements difficult to assure in large-scale programs of aerial spraying.

Outbreaks of the Douglas-fir tussock moth typically are explosive in nature, with populations rising from unnoticed to tree-killing numbers in 1–2 years (Wickman *et al.,* 1971). The outbreaks usually are smaller in area and shorter in duration than those of the budworm. The current tussock moth outbreak in Oregon, Washington, and Idaho appears to be an exception; it is the most extensive and damaging yet recorded.

We do not have sufficient information on the underlying factors affecting tussock moth abundance and distribution to anticipate when and where it will appear in epidemic numbers. Nor can we predict how rapidly it will build up, how far it will spread, or when the population will decrease after it has been detected in a given area. We do know that a disease of the insect caused by a nucleopolyhedrosis virus is often associated with collapsing populations; recent field experiments with this virus indicate that it may be useful for direct suppression of populations of the tussock moth before they reach tree-killing numbers (Thompson and Markin, 1973). *Bacillus thuringiensis* and several candidate chemical insecticides may be effective in suppressing tussock moth populations. However, no insecticide is presently registered for use against this insect.

The most critical needs for both the budworm and the tussock moth are: (1) more information on their population dynamics to provide a basis for prediction of outbreaks and selection of management strategies, (2) a monitoring procedure that will permit earlier detection of population increases, especially in old-growth stands, and (3) better information on potential losses in old-growth stands. With these three additions to present knowledge, more rational decisions can be made as to whether direct control action or no action other than salvage of killed timber is the better solution.

Poria root-rot. This disease is caused by a fungus, *Poria weirii,* which attacks many conifers in the Pacific Coast states; but Douglas-fir is the most susceptible. The disease develops in small infection centers that may become half a hectare or more in area. Infection centers are much more abundant in certain localities than in others, but the cause of this variation is unknown. Frequently the trees are killed or their roots decayed so completely that they are toppled by wind. *Poria weirii* may exist for many years in buried remnants of diseased trees or invade healthy stands by occasional basidiospore infections. From these colonized materials the disease can spread to adjacent younger trees and stands through root contacts.

Losses due to this disease can be decreased by certain management practices (Childs and Nelson, 1971): (1) early cutting of stands that show the greatest incidence of infection centers and deferring cutting in less infected stands, (2) salvaging standing trees in infection centers to prevent subsequent loss, and (3) selective cutting to favor less susceptible tree species.

In replanting infected areas, western red cedar and red alder have been advocated both in mixed and in pure stands, western red cedar, because it is less susceptible to Poria root rot than Douglas-fir (Childs and Nelson, 1971), and alder because nitrogen-fixing nodules associated with its root system increase the amount of nitrate-nitrogen in soils (Bollen and Lu, 1968). Nitrate-nitrogen cannot be utilized by *Poria weirii* but is utilized by many competing wood-inhabiting fungi (Li *et al.*, 1967). Thus, accumulation of nitrate-nitrogen may decrease the inoculum potential of the fungus by favoring the development of organisms antagonistic to it in roots and other woody debris (Nelson, 1969). The length of time and density of alder necessary to achieve these two beneficial effects has yet to be determined. This possible disease management strategy well illustrates the need for sound and thorough investigation of the complex biological and chemical interactions within forest ecosystems when developing integrated management strategies for forest pests.

Much more information is needed about the biology and economics of all the above pests. This information must be developed before effective and efficient management strategies can be developed for each pest individually. Thus, an integrated program, considering all significant pests simultaneously, is still a very long-range goal.

The Southern Pine Forest

We will now examine several major insect and disease problems in another type of forest which is also geared to timber production, but where protection involves much more than keeping the older trees alive. The young pine forests of the southern United States are an excellent example. Here, the primary objective is to produce maximum yields of wood fiber per unit area of land in a given period of time. Rotation times may be as short as 15–20 years for pulpwood and 40–80 years for sawtimber or veneer bolts. Protection against destructive insects and diseases must be provided from the time the trees are planted or naturally regenerated to time of harvest. This is essential to ensure rapid growth and avoid extending rotation times and other disruptions in management schedules and operations.

The southern pine forests contain about one-sixth of the total softwood growing stock in the United States (U.S. Forest Service, 1973). In 1970, they produced one-fourth of the nation's supply of softwood lumber and plywood, and three-fourths of its needs for pulpwood. The major pine species of the region include loblolly, slash, shortleaf, and longleaf pines. Virtual elimination of heartrot in all the southern pines from an average of 30% loss by volume in the early 1900s to less than 1% loss in the 1970s, provides an outstanding example of success in minimizing disease loss through effective forest management. This was achieved mainly by decreasing rotation ages below the age threshold at which heartrot becomes important.

The major insect and disease pests of southern pine forests are the southern pine beetle (*Dendroctonus frontalis*), fusiform rust caused by *Cronartium fusiforme,* littleleaf disease—a disease complex caused by poor soil conditions and the feeder-root fungus, *Phytophthora cinnamomi,* and Annosus root rot caused by *Fomes annosus*. Individually and jointly, these four pests have a substantial impact on the productivity and value of the southern pine resource.

The Southern Pine Beetle. This beetle is the most destructive insect enemy of the southern pines. Localized infestations occur almost every year from eastern Texas to Pennsylvania. The current outbreak, which has been more or less persistent since 1957 (Thatcher, 1960; Coulson *et al.,* 1972), increased greatly in intensity in 1972 and 1973. Through cooperative efforts of private timberland owners, state agencies, and the U.S. Forest Service, approximately 413,000 cubic meters (172 million board feet) of sawtimber and 1.7 million cubic meters (481,000 cords) of beetle-attacked pulpwood were salvaged in 1973 (U.S. Forest Service, 1975). This was about three-fourths of the total volume of timber killed.

Variation in the life history, seasonal behavior, and occurrence of natural controls of the beetle, make it very expensive to assess the status of beetle populations and to predict changes in their numbers and distribution from year to year. The beetle produces 3–4 generations per year in the northerly and 5–7 generations per year in southerly parts of its range (Thatcher, 1960). From Georgia to eastern Texas the beetle is active in all months of the year; during the summer, the interval between attack and emergence of a new brood may be 30 days or less.

Some general patterns in the development of outbreaks provide clues to improved silvicultural practices that may alter beetle populations. Outbreaks appear most frequently where trees are growing on poorly drained and infertile soils, where there has been prolonged moisture stress from drought, in dense stands where tree growth is continuously suppressed, or where there is considerable damage to stands from harvesting or other silvicultural operations (Moore and Thatcher, 1973). Lightning-struck trees also provide foci for beetle attacks and subsequent buildups. Thus, the hazard of losses due to the beetle can be decreased by: (1) careful selection of sites for planting, (2) maintenance of a density of growing stock that is within the capacity of the site to support vigorous tree growth, (3) prompt removal of lightning-struck trees, and (4) harvesting operations designed to avoid damage to the residual trees.

The southern pine beetle also has numerous natural enemies (Thatcher, 1960; Coulson *et al.,* 1972). Various parasites, vertebrate and invertebrate predators, and pathogens have been reported to decrease beetle populations.

But no practical means of augmenting or enhancing the effects of these natural enemies has yet been developed.

The treatment of infested trees with insecticides, notably Lindane, has been shown to be effective when properly applied before the beetles emerge from the tree (Thatcher, 1960). Such treatment requires very early detection, in-place treatments of the entire tree bole, or felling the trees and then rolling each log over completely so that the chemical can be applied to the entire bark surface. There is some evidence that this treatment also kills some of the primary parasites and predators of the beetle; thus its long-term effectiveness must be questioned.

Recent studies with herbicidal compounds containing cacodylic acid and attractants containing pheromones and certain host-produced chemicals indicate that these materials may be useful in manipulating and/or suppressing beetle populations (Coulson *et al.*, 1972).

The U.S. Forest Service and university cooperators * are intensifying efforts to find ways and means of utilizing these different methods and approaches effectively in an integrated management system. Specific guidelines are needed for the resource manager to determine when, where, and how each method should be applied and, more broadly, the optimal strategies for their combined use to maintain forest productivity on a long-term basis. More intensive monitoring of beetle populations and pine stand conditions is an essential prerequisite for efficient management of the southern pine beetle. Procedures have been developed for detecting and evaluating beetle infestations from the air and ground (U.S. Forest Service, 1970); but these must be greatly improved and intensified if the resource manager is to obtain information about population buildups early enough for efficient planning and scheduling of control operations.

Fusiform Rust. During the past 20 years, fusiform rust (caused by *Cronartium fusiforme*) has become the most economically damaging disease of the southern pines; current losses are estimated to be about $28 million annually throughout the South (Powers *et al.*, 1974). Slash and loblolly pines are highly susceptible, while longleaf pine is highly resistant and shortleaf pine is essentially immune to this disease. Affected trees frequently are deformed, predisposed to wind breakage, or killed by the pathogen. The disease is most severe in plantations less than 15 years old. Recently completed surveys indicate that the disease occurs throughout the natural range of the southern pines but is most damaging in a zone extending from coastal South Carolina, Georgia, and north Florida across Alabama, Mississippi, and Louisiana

* With additional support from the project sponsored by the National Science Foundation–Environmental Protection Agency "Principles, Strategies, and Tactics of Pest Population Regulation and Control in Major Crop Ecosystems" (NSF Grant No. GB-34718, C. B. Huffaker, University of California, Berkeley, Director).

(Phelps, 1973). Stands with 90% or more basal stem infections are known in all of these states; such stands frequently are not profitable even for salvage cuttings.

The disease has been known in the South since the late 1800s. The causal fungus develops alternately in pine and various species of oak. Oaks are rarely harmed by the disease which cannot spread from pine to pine. The fungus infects growing shoots and needles of pine, and then develops in synchrony with the cambium of branches and stems which are stimulated to form characteristic spindle-shaped galls and cankers. These are invaded by insects, wood-staining fungi, and decay fungi.

The marked increase in prevalence of fusiform rust during the past 20 years has been attributed primarily to the following activities of man: (1) widespread planting of infected nursery stock (Hodges, 1962), (2) establishment of even-aged monocultures of susceptible slash and loblolly pines over large areas of land that previously supported rust-resistant longleaf pine and frequently included sites to which these species are not well adapted (Siggers and Lindgren, 1947), (3) more widespread planting of slash pine than of loblolly pine, (4) increased populations of alternate (oak) hosts due to suppression of wild fire, and (5) intensive management practices such as site preparation (Miller, 1972) and forest fertilization (Dinus and Schmidtling, 1971), which apparently predispose planted seedlings to increased susceptibility.

Meteorological conditions favoring infection have been carefully described (Snow, 1968*a,b;* Snow and Froehlich, 1968; Snow *et al.*, 1968). Substantial genetic variation in resistance has been found in both loblolly and slash pines (Kinloch and Stonecypher, 1969; Stonecypher *et al.*, 1973). Rapid procedures for screening seedlings for rust resistance have been developed (Miller, 1970; Matthews and Rowan, 1972; Powers, 1974).

Establishment of forest nurseries in areas of low rust hazard and applications of fungicides have provided very satisfactory and economical methods for control of fusiform rust in forest nurseries; this has substantially decreased losses attributable to planting of infected nursery stock (Hodges, 1962).

Fusiform rust usually is less damaging in naturally regenerated than in planted stands. In infected natural stands, the only feasible methods of limiting losses due to fusiform rust are: (1) pruning of branch galls before the disease enters the tree stem, and (2) selective removal of trees with stem galls before the trees die or are broken by wind. In areas to be regenerated naturally, disease resistance can be increased by selecting as seed trees, rust resistant shortleaf and longleaf pines and/or disease-free loblolly or slash pines.

Planting of seedlings genetically resistant to fusiform rust is the most promising method of minimizing losses due to this disease in planted stands. Longleaf pine is highly resistant to fusiform rust and shortleaf pine is essentially immune. Loblolly pine is more tolerant of fusiform rust than slash pine—

capable of producing a merchantable tree despite infection (Cole, 1975). Wells (1969) has shown that substantial geographical variation in resistance to fusiform rust exists in loblolly pine. In the Gulf Coast states he has advocated planting seedlings produced from seed collected in Livingston Parish and nearby areas of Louisiana.

Beginning in the late 1960s, special orchards of highly rust-resistant selections of loblolly and slash pines were established by the U.S. Forest Service and the Cooperative Tree Improvement Programs at North Carolina State University and the University of Florida. By 1980 these orchards, and others developed subsequently, should be producing enough seed to satisfy the demand for resistant seedlings in the area of greatest rust losses mentioned above. To further speed up the process of selecting resistant genotypes of loblolly and slash pines, the U.S. Forest Service established a central Fusiform Rust Resistance Testing Center at Asheville, N.C. in 1971 (Lewis and Cowling, 1970). It is now operational and is being used to test hundreds of families of select trees per year (Laird *et al.*, 1973).

Present chemical spray formulations and schedules for forest nurseries are governed primarily by the rule of thumb ''spray twice a week and after every rain.'' Unverified trust that sprays are eliminating seedling infections has apparently led to elimination of inspections to cull infected seedlings before outplanting. Economic analyses of gains from improvement in rust resistance are now being made (Porterfield, 1973). Substantial geographic and genetic variation in virulence of the rust pathogen is known but its implications for breeding programs and deployment of resistant families have yet to be determined. A recent rust-incidence survey has identified general areas of high and low disease hazard (Phelps, 1973). Research to evaluate the influence of various site factors such as nutrient availability, amount of oak, average leaf wetness, etc., are now underway. Oak incidence has been shown to be highly correlated with disease incidence in north Florida, Georgia, and South Carolina. This is encouraging in that estimates of disease hazard throughout the region may soon make it possible to deploy disease-resistant planting stock in sites of maximum disease hazard. Studies of the impact of fusiform rust on pulping properties of infected trees have shown that economic losses in pulpwood rarely exceed a few cents per cubic meter so long as the affected trees remain alive (Veal *et al.*, 1974). Based on these and other economic data, analyses are being completed to determine optimum harvesting times for stands with various amounts of stem infection. In short, a predictive system of management for fusiform rust should soon be available for integration into general resource management plans for the southern pines.

Littleleaf Disease. This disease has caused extensive mortality, stagnation of stands, and generally poor growth of shortleaf pine in large areas of the Piedmont region of central Virginia, North Carolina, South Carolina, Georgia,

Alabama, and Mississippi, where heavy clay soils were badly abused, eroded, and finally abandoned to agriculture. In some areas, littleleaf disease has essentially eliminated commercial production of shortleaf pine; in other areas within the above region the disease has remained unimportant to this day. Hepting (1949) estimated total losses in value of shortleaf pine to be about $5 million annually.

Littleleaf disease is one of the most complicated diseases of forest trees. At the same time, however, it is one of the best understood in terms of how the disease is induced, how it affects the host, and how it can be managed.

Littleleaf disease is caused by a deficiency of nitrogen, which results from death or impairment of feeder root function caused by *Phytophthora cinnamomi*. This killing or impairment of feeder roots occurs only under certain essential predisposing soil conditions. These conditions include badly eroded, shallow soils of high clay content, soils poorly drained or impervious to water, those poorly aerated, deficient in organic matter, and subject to periodic moisture stress (Campbell and Copeland, 1954). The fungus pathogen occurs in pine forests throughout the southern United States; but the disease is limited in distribution to soils that frequently remain high in moisture and low in aeration. The moisture is needed to serve as a medium for movement of the motile swarm-spores of the fungus; poor aeration apparently limits the capacity of the host to form new feeder roots or of existing feeder roots to satisfy the demand of the tree for nitrogen. The disease rarely develops in stands less than about 20 years of age. When it does develop, it is manifest as a classic nutrient- and water deficiency syndrome. The tree produces short yellowish needles (hence the name littleleaf), small annual increments of diameter and height growth, small cones, and small seeds of poor germinability. In addition, the nitrogen content of foliage and the amounts of reserve carbohydrates stored in the roots decline. The plants also experience periodic water stress.

Management of littleleaf disease is accomplished by an integration of methods that involves site hazard evaluation, fertilization treatments, silvicultural treatments, and use of resistant species of pine for regeneration of affected sites. In addition, site amelioration, selecting and breeding of resistant genotypes of shortleaf pine, and controlled manipulation of protective mycorrhizae may be feasible in the future.

In affected trees of high value, such as in parks, residential areas, and roadside locations, fertilization frequently will alleviate the symptoms and prevent mortality. Managers of commercial forest, on the other hand, should be guided by the following principles (Hepting, 1949): (1) stands usually attain an age of 30–50 years before they begin to break up from littleleaf; (2) littleleaf trees usually die 6 years after onset of typical symptoms, and 2–3 years after advanced symptoms; (3) where littleleaf is developing in stands about 30 years of age, repeated light cuts or a few heavy cuts may remove most diseased trees before the stand is 50 years old; (4) since affected trees generally reach a diameter of 20 cm before they die, severe littleleaf areas (soil rating scale <50 in Table II) are probably better suited for pulpwood production than for sawtimber.

Site-hazard evaluation is accomplished with a simple soil-rating scale based on four criteria that are readily measured in the field (Campbell and Copeland, 1954) (see Table II). If the rating scale is less than 50, littleleaf disease will be severe and alternative species such as loblolly pine, longleaf pine, Virginia pine, or hardwoods should be favored. If the rating scale is above 75, littleleaf disease will not develop and shortleaf pine can be used. If the rating scale is between 50 and 75, the hazard of disease will be moderate and either alternative species or very careful matching of basal area with carrying capacity for shortleaf pine will need to be followed. Specific cutting guidelines for the timing and intensity of cutting to improve littleleaf stands are given by Campbell and Copeland (1954).

Several soil amelioration treatments also are becoming economically feasible. These include: (1) improvement of internal drainage by breaking up the water-impervious layers that underlay most severe littleleaf sites, and (2) plowing in or surface applications of various organic amendments such as sewage sludge and bark which has been augmented with nitrogen fertilizers. Genetic variation in resistance to littleleaf disease has been noted in many affected stands of shortleaf pine and progeny of these select trees are being evaluated for their potential in regeneration of severe littleleaf sites. Protection of feeder roots by certain mycorrhizae has been demonstrated by Marx and Davey (1969). Inoculation of sites with ecologically adapted mycorrhizae also is being tested as a possible future technique of disease management.

The major weak links in the above recommendations for littleleaf disease

TABLE II. Soil Rating Values for Site-Hazard Evaluation of Littleleaf Disease

Soil characteristics	Value	Soil characteristics	Value
Erosion		Depth to zone of greatly	
Slight	40	increased permeability	
Moderate	30	60–90 cm	15
Severe	20	45–58 cm	12
Rough gullied	10	30–44 cm	9
Soil consistence (when moist)		14–28 cm	6
Very friable	32	0–12 cm	3
Friable	24	Subsoil mottling	
Firm	16	None	13
Very firm	8	Slight	9
Extremely firm	0	Moderate	5
		Strong	1

are (1) lack of an aggressive extension program to persuade forest managers to take full advantage of the available knowledge, (2) lack of integration with control recommendations for other pests such as Annosus root rot and the southern pine beetle.

Annosus Root-Rot. This root disease is caused by *Fomes annosus,* a wood-decay fungus which attacks all conifers in the southern United States. Its primary effect is killing of trees in plantations and natural stands after thinning. The causal organism usually invades the cut surface of stumps and then spreads through root contacts to adjacent trees. A survey in 1960 (Powers and Verrall, 1962) indicated average losses due to mortality and wind-throw of about 3% of the trees in loblolly and slash pine plantations throughout the South (although losses as great as 30% were reported in individual plantations).

Annosus root rot is one of the most thoroughly researched of all forest pests; a bibliography of references published in the decade between 1960 and 1970 contained 594 citations (Hodges *et al.,* 1971). Although the disease was once considered to have substantial epidemic potential, more recent research studies now indicate that the disease is not likely to become epidemic (Kuhlman *et al.*, 1975). However, forester managers will need to take precautions to avoid needlessly heavy losses in areas of high disease hazard.

Guidelines for management of *Fomes annosus* in the southern United States have recently been developed by Kuhlman *et al.* (1976), and the U.S. Forest Service research project on this disease has been terminated. Thus, the primary responsibility for this disease is now shifting from forest pathologists to forest managers, with emphasis on the integration of recommendations for preventing or decreasing losses by this disease with general resource management and utilization practices.

The management guidelines for the disease are based on the use of a site-hazard evaluation system to determine where to apply the following preventive treatments: (1) use of wide spacing to decrease the necessity for early thinning of plantations, (2) use of stump protectants such as borax to prevent infection, (3) use of other fungi, such as *Peniophora gigantea* to compete with *F. annosus* in cut stumps and in roots, (4) restricting thinnings to low-risk seasons in the year (e.g., thinning in May–August south of 34°N latitude involves little risk of stump infection), (5) decreasing the number of thinnings per rotation, (6) use of harvesting procedures designed to remove as much as possible of the stump and tap rot, (7) regeneration by natural or direct seeding as opposed to planting of seedlings, and (8) use of longleaf pine, which appears to be somewhat more resistant to Annosus root rot than loblolly, slash, or shortleaf pines.

The present limitations of these guidelines as a pest management plan are (1) the need for a more simple and universally applicable quantitative procedure for disease-hazard evaluation in the field, (2) lack of adequate cost-benefit data for decisions among the eight preventive treatments listed above, (3) the need for more aggressive extension of knowledge about the value and utility of the above guidelines, and (4) lack of integration of these control guidelines with considerations for bark beetles and other pests of the southern pines.

Conclusions. In the case of littleleaf disease and Annosus root-rot, present management guidelines for southern pines are approaching a reliable and adequate pest-management system. In the case of fusiform rust and especially in the case of the southern pine beetle, much remains to be learned before adequate management guidelines can be developed and applied effectively.

The Northeastern Hardwood Forest

For contrast, we will now examine a type of forest in which timber production is of secondary importance and protection is mainly directed at optimizing other values. The deciduous forests of the northeastern United States provide a good example. These forests are even more diverse ecologically and are subject to a wider range of depredations by many different insects and diseases than the old-growth Douglas-fir and southern pine forests. The economics of managing and protecting northeastern hardwood forests are greatly complicated by the wide range of values placed on them by their owners and the public at large.

Despite its sprawling urban centers and heavy concentrations of industry, the northeastern United States is the most heavily forested part of the country (U.S. Department of Interior, 1970). Four-fifths of the total land area in the New England states and half that in the Middle Atlantic states is covered by forests. Some valuable wood products are derived from these forests, but other uses such as watershed protection, wildlife habitat, recreation, and aesthetics are much more important to the general welfare and economy of the region. Much of the public concern over the gypsy moth and other defoliating insects and the widespread diebacks and declines of some major hardwood species relates to their effects on the appearance and health of the forest as a backdrop for suburban living and tourism.

Defoliation by the gypsy moth extended over 0.6 million hectares in nine states in 1971 and 1972 and an additional 0.8 million hectares in 1973 (U.S. Forest Service, 1974). Thousands of additional hectares were defoliated in this period by other insects such as the forest tent caterpillar (*Malacosoma disstria*), elm spanworm (*Ennomos subsignarius*), fall cankerworm (*Alsophila pometaria*), oak leaf rollers and tiers (*Archips* sp. and *Croesia* sp.), and large aspen tortrix (*Choristoneura conflictanta*). Collectively and cumulatively, this defoliator complex poses serious problems in the complicated economic, social, and political decision-making structure of forest pest control in the northeastern states.

The gypsy moth is the only forest insect in the United States that is still under a federal quarantine. In the border areas of its present distribution it is accorded all of the attention usually given an unwanted exotic pest—massive efforts to exclude or eliminate it. By contrast, in the generally infested parts of New England, New York, New Jersey, and Pennsylvania, where it is well established, it is considered and treated as one of the native insect complex.

Defoliation surveys of the forested areas of the northeastern states are conducted more or less regularly by state forestry or pest control agencies, mostly in cooperation with the U.S. Forest Service. Additional reports of insect activity are received from private landowners, woods workers, campers, hikers, and other individuals who have observed some defoliation or other damage in the forest. With the large number of people working in or traveling through the forested areas, many of them having some sensitivity to the appearance of the trees, rarely does an infestation go long undetected.

More options for direct and indirect suppression of damaging populations are available for the gypsy moth than for any of the other forest insects in the northeastern United States. Chemical insecticides, parasite and predator releases, and silvicultural manipulations of host stands have all been applied with various degrees of success. Two commercial formulations of *Baccillus thuringiensis* have recently been registered for use against the gypsy moth. Interestingly, the efficacy of these formulations was established primarily on the basis of foliage protection, rather than population suppression. A nucleopolyhedrosis virus, which is an important regulating factor in natural populations of the insect, has been field tested and subjected to detailed safety evaluation. It may be available in the near future for direct control. Research on the sex attractant, disparlure, antifeeding compounds, and release of sterile males and genetically deleterious strains has indicated that in certain circumstances all might be useful in suppressing populations of the gypsy moth and/or protecting their host trees. Most of these approaches may also be applicable to the other major defoliating insects, but research on them is less advanced. For these pests, chemical insecticides still are relied upon to provide protection.

In managing this complex of defoliating insects, the greatest need is a consistent, objective basis for deciding if, when, and where to take action. More basic information is needed to assess the impacts of the insects on the intangible as well as the tangible values of the hardwood forest. New analytical techniques are needed to link the ecological, economic, and sociopolitical factors involved, and to determine the values to be protected or gained by different control strategies or actions. Valid, defensible cost-benefit evaluations can then be made as a basis for decision making.

A comprehensive research and development program by the U.S. Department of Agriculture and cooperating state and private organizations is aimed at developing a total management system for the gypsy moth (U.S. Department of

Agriculture, 1971). This program, initially funded at $1.8 million a year for a 5-year period, indicates the scale and intensity of effort needed to develop an integrated management system for the entire defoliator complex.

In addition to the above complex of defoliating insects, northeastern hardwoods also suffer from dieback and decline diseases that have become increasingly prevalent in stands and roadside plantings of birch, ash, oak, maple, beech, and sweetgum. All of these species of trees show a complex series of common symptoms. Houston (1967, 1973) has presented an excellent description of the symptomology, etiology, and possible management of these diseases. Typically the uppermost branches of the trees die back from the top. Leaves are small and chlorotic, bunched only at the tips of branches, or absent entirely. Sometimes an unusual number of branch sprouts arise from adventitious buds. These various symptoms are due to one, or usually several, stresses resulting from insect defoliation, drought, excessive soil temperature, road salt, air pollutants, or late-spring frosts in combination with many secondary fungus pathogens.

When the initial stress is biotic in origin, such as insect defoliation, it may be possible to prevent dieback and decline of the trees by direct control of the predisposing stress factor. When the initial stress is abiotic in origin, however, the problem is more difficult. The causes of this stress must be identified and effective action taken to ameliorate it, if possible. Most important, decisions about all possible management procedures should be based on the same ecological and socioeconomic criteria as for the defoliating insects. The management programs for these insects and diseases should be directly coordinated.

Summary and Recommendations

An adequate program of forest pest management is an obvious necessity for wise forest resource management. In this paper we have described forests as dynamic ecosystems that provide renewable resources for many human needs. The ecological diversity of forests; their particular patterns of ownership; and unique economic, legal, and social values and constraints; add to the complexities of developing adequately comprehensive and efficient systems for management of forest pests. All elements of forest resource use must be considered in relation to the interplay of each insect and disease as well as all other destructive agents. Each one of these various elements must be put together into an operational model for planning and decision making. What is needed is a system for unit-area management that takes into account all silvicultural, socioeconomic, and pest factors that prevail, or are likely to prevail, within that area in the long run as well as in the short run. The basic components of forest pest management systems and essential steps in their development are outlined earlier in this paper with examples of current practice and future needs in three major types of forest.

Much greater communication and cooperation between forest resource managers and specialists in the various sciences basic to forestry need to be achieved if efficient use is to be made of developing conceptual and practical knowledge. The following specific recommendations are offered in the hope of accelerating the achievement of these goals:

1. *Pest-caused loss and damage appraisal.* More intensive and efficient methods are needed for the timely detection and economic assessment of losses due to forest insects and diseases. Such information is needed especially as guides for development of rational research and extension priorities. Greater cooperation between forest entomologists and pathologists and forest survey personnel would be a valuable initial step in this direction. Adequate loss-assessment information and realistic impact analyses will require inputs by forest economists and mensurationists as well as entomologists and pathologists.

2. *Pest-resistance research.* Genetically controlled variation to many forest pests exists in many forest tree species. This variation offers the potential for selection and breeding of resistant genotypes. Exploitation of available genes for resistance will require greater cooperation between forest geneticists and forest pathologists and entomologists.

3. *Pest-hazard evaluation.* Forest soil and climatic conditions have major influences on the population dynamics, epidemiology, and infection potential of various disease and insect pests of forest trees. Greater communication and cooperation among specialists in silviculture, forest soils, meteorology, air pollution effects on vegetation, entomology, and pathology will provide more effective integration of knowledge in these now distinct subdisciplines of forestry.

4. *Ecosystem analysis of pest effects.* Traditionally, forest entomologists have been concerned mainly with specific insects and only secondarily with the physiology and ecology of their host trees. Also, many forest pathologists have pursued studies of tree diseases without an adequate regard for the whole forest ecosystem in which these diseases operate. This has led to some short-sightedness which could be avoided by greater cooperation with silviculturalists and forest ecologists and by adopting a more ecosystem-oriented view in the pursuit of specific pest-research objectives.

5. *Quantification of nontimber values.* No need is more urgent in forestry today than more effective means to quantify the value of forests as sources of water, recreation, wildlife, and as an aesthetic backdrop for our increasingly urban society. Innovative approaches, from whatever source, would be welcomed by forest scientists and forest managers of all types.

6. *Communication with forest managers.* The ultimate utility of forest

pest management depends on its effective communication to forest resource managers who can put the concept into practice.

7. *Utilization of pest-damaged timber*. In many cases, a substantial potential exists to decrease losses due to insects and diseases by developing new uses and markets for pest-damaged timber. Expanded contacts between pest control personnel, wood and paper technologists, and wood-products marketing and development specialists will be helpful.

8. *Forest pest advisory services*. In many cases, forest pest control personnel have done a better job of discovering new knowledge about tree diseases and insects than they have of communicating knowledge in their special field to forest managers and land owners who could benefit from their research. Extension programs have been mainly passive, that is, they have been responsive to requests for advisory service rather than aggressive in anticipating educational needs. Overt programs designed to deliver information *to* forest managers as well as to develop information *for* managers would do much to increase the effectiveness of present pest advisory services. Attractive and persuasive "packaging" of research information also would be helpful in increasing the benefits from forest disease and insect research.

ACKNOWLEDGMENTS

The authors acknowledge with appreciation the following persons who offered valuable suggestions and comments on portions of this manuscript: G. H. Hepting, H. R. Powers, Jr., D. M. Knutson, J. R. Parmeter, R. J. Dinus, J. L. Stewart, and E. E. Nelson.

Literature Cited

Bollen, W. B., and Lu, K. C., 1968, Nitrogen transformation in soils beneath red alder and conifers, *in: Biology of Alder* (J. M. Trappe, J. F. Franklin, R. F. Tarrant, and G. M. Hansen, eds.), Pacific Northwest Forest and Range Experiment Station, Portland, Ore.

Boyce, J. S., 1932, Decay and other losses in Douglas fir in Western Oregon and Washington, *U.S. D. A. Tech. Bull.* **286.** 60 pp.

Campbell, W. A., and Copeland, O. L., 1954, Littleleaf disease of shortleaf and loblolly pines, *U.S. D. A. Circ.* **940.** 41 pp.

Carolin, V. M., Jr. and Orr, P. W., 1972, Western spruce budworm, U.S. Department of Agriculture, Forest Service, Forest Pest Leaflet 53, Washington, D.C. 8 pp.

Childs, T. W., and Nelson, E. E., 1971, Laminated root rot of Douglas-fir, U.S. Department of Agriculture, Forest Service, Forest Pest Leaflet No. 48, Washington, D.C. 7 pp.

Cole, D. E., 1975, Comparisons within and between populations of planted slash and loblolly pine, Georgia Forest Research Paper No. 81, Georgia Forest Research Council, Macon. 13 pp.

Coulson, R. N., Payne, T. L., Coster, J. E., and Houseweart, M. W., 1972, The southern pine beetle *Dendroctonus frontalis* Zimm. (Coleoptera: Scolytidae) 1961–1971, Texas Forest Service Publication 198, College Station. 38 pp.

Dinus, R. J., 1974, Knowledge about natural ecosystems as a guide to disease control in managed forests, *Proc. Amer. Phytopath. Soc.* **1:**184–190.

Dinus, R. J., and Schmidtling, R. C., 1971, Fusiform rust in loblolly and slash pines after cultivation and fertilization, U.S. Department of Agriculture, Forest Service Research Paper 50-68, New Orleans, La. 10 pp.

Duerr, W. A., Teeguarden, D. E., Guttenberg, S., and Christiansen, N. B., 1974, *Forest Resource Management: Decision-Making Principles and Cases,* Oregon State University Bookstores, Corvallis. 500+ pp.

Furniss, M. M., and Orr, P. W., 1970, Douglas-fir beetle, U.S. Department of Agriculture, Forest Service, Forest Pest Leaflet No. 5, Washington, D.C. 4 pp.

Furniss, M. M., Kline, L. N., Schmitz, R. F., and Rudinsky, J. A., 1972, Tests of three pheromones to induce or disrupt aggregation of Douglas-fir beetles (Coleoptera: Scolytidae) on live trees, *Ann. Entomol. Soc. Am.* **65:**1227–1232.

Hepting, G. H., 1949, Managing pines in littleleaf areas, *For. Farmer* **8(11):**7–10.

Hodges, C. S., Jr., 1962, Diseases in southeastern forest nurseries and their control, Southeastern Forest Experiment Station Paper No. 142, Asheville, N.C. 16 pp.

Hodges, C. S., Jr., Koenigs, J. W., Kuhlman, E. G., and Ross, F. W., 1971 *Fomes annosus:* A bibliography with subject index—1960–1970, U.S. Department of Agriculture Forest Service Research Paper SE-84, Asheville, N.C. 75 pp.

Houston, D. R., 1967, The dieback and decline of northeastern hardwoods, *Trees* **28:**12–14.

Houston, D. R., 1973, Diebacks and declines: Diseases initiated by stress, including defoliation, *Proc. Intern. Shade Tree Conf.* **49:**73–76.

Kinloch, B. B., and Stonecypher, R. W., 1969, Genetic variation in susceptibility to fusiform rust in seedlings from a wild population of loblolly pine, *Phytopathology* **59:**1246–1255.

Knopf, J. A. E., and Pitman, G. P., 1972, Aggregation pheromone for manipulation of the Douglas-fir beetle, *J. Econ. Entomol.* **65:**723–726.

Kuhlman, E. G., Hodges, C. S., Jr., and Froehlich, R., 1976, A prescription for management of *Fomes annosus* in the southern United States, *J. Forestry* (In press).

Laird, P. O., Knighten, J. L., and Wolfe, R. L., 1973, An evaluation of a method for use in determining relative fusiform rust resistance in loblolly and slash pine families, Forest Tree Resistance Testing Facility Report No. 1, U.S. Department of Agriculture Forest Service, State and Private Forestry, Asheville, N.C. 15 pp.

Lewis, R. A., and Cowling, E. B., 1970, Proposal for creation of a southern pine rust resistance testing center, Department of Plant Pathology, North Carolina State University, Raleigh. 12 pp.

Li, C. Y., Lu, K. C., Trappe, J. M., and Bollen, W. B., 1967, Selective nitrogen assimilation by *Porra weirii, Nature* (*Lond.*) **213:**841.

Marty, R., 1966, Economic guides for blister-rust control in the East, U.S. Department of Agriculture, Forest Service, Research Paper NE-45, Upper Darby, Pa. 14 pp.

Marty, R., and Mott, D., 1964, Evaluating and scheduling white-pine weevil control in the Northeast, U.S. Department of Agriculture, Forest Service, Research Paper NE-19, Upper Darby, Pa. 56 pp.

Marx, D. H., and Davey, C. B., 1969, The influence of ectotrophic mycorrhizal fungi on the resistance of pine roots to pathogenic infections. IV. Resistance of naturally occurring mycorrhizae to infections by *Phytophthora cinnamomi, Phytopathology* **59:**559–565.

Matthews, F. R., and Rowan, S. J., 1972, An improved method for large-scale inoculations of pine and oak with *Cronartium fusiforme, Plant Dis. Reptr.* **56:**9321–934.

Miller, T., 1970, Inoculation of slash pine seedlings with stored basidiospores of *Cronartium fusiforme, Phytopathology* **60:**1773–1774.

Miller, T., 1972, Influence of site preparation and spacing on the incidence of fusiform rust in planted slash pine, *For. Sci.* **18:**70–75.

Moore, G. E., and Thatcher, R. C., 1973, Epidemic and endermic populations of the southern pine beetle, U.S. Department of Agriculture Forest Service, Research Paper SE-111, Asheville, N.C. 11 pp.

Nelson, E. E., 1969, Occurrence of fungi antagonistic to *Poria weirii* on a Douglas-fir soil in Western Oregon, *For. Sci.* **15:**49–54.

Phelps, W. R., 1973, Survey of fusiform rust in southern pine plantations, U.S. Forest Service, State and Private Forestry, Atlanta, Ga. 73 pp.

Porterfield, R. L., 1973, Predicted and potential gains from tree improvement programs—A goal programming analysis of program efficiency, Ph.D. dissertation, Yale University, New Haven, Conn. 240 pp.

Powers, H. R., 1974, Breakthrough in testing for fusiform rust resistance, *For. Farmer* **33:**7–8.

Powers, H. R., Jr., and Verrall, A. F., 1962, A closer look at *Fomes annosus, For. Farmer* **21:**8–9, 16–17.

Powers, H. R., Jr., McClure, J. P., Knight, A. A., and Dutrow, G. F., 1974, Incidence of fusiform rust in the Southeastern United States as determined from forest survey data, *J. Forestry* **73:**398–401.

Rudinsky, J. A., 1968, Pheromone-mask by the female *Dendroctonus pseudotsugae* Hopk., an attraction regulator, *Pan-Pacific Entomol.* **44:**248–250.

Siggers, P. V., and Lindgren, R. M., 1947, An old disease—A new problem, *South Lumberman* **175(2201):**172–175.

Snow, G. A., 1968*a,* Time required for infection of pine by *Cronartium fusiforme* as affected by suboptimal temperatures and preconditioning of teliospores, *Phytopathology* **58:**1547–1550.

Snow, G. A., 1968*b,* Basidiospore production by *Cronartium fusiforme* as affected by suboptimal temperatures and preconditioning of teliospores, *Phytopathology* **58:**1541–1546.

Snow, G. A., and Froehlich, R. C., 1968, Daily and seasonal dispersal of basidiospores of *Cronartium fusiforme, Phytopathology* **58:**1532–1536.

Snow, G. A., Froehlich, R. C., and Popham, T. W., 1968, Weather conditions determining infection of slash pines by *Cronartium fusiforme, Phytopathology* **58:**1537–1540.

Society of American Foresters, 1964, Forest cover types of North America, Washington, D.C. 67 pp.

Stonecypher, R. W., Zobel, B. J., and Blair, R. L., 1973, Inheritance patterns of loblolly pines from a nonselected natural population, *N.C. Agr. Exp. Sta. Tech. Bull.* **220.** 60 pp.

Thatcher, R. C., 1960, Bark beetles affecting southern pines: A review of current knowledge, U.S. Department of Agriculture, Forest Service, South Forest Experiment Station, Occasional Paper 180. New Orleans, La. 25 pp.

Thompson, C. G., and Markin, G. P., Office report, Pacific Northwest Forest and Range Experiment Station, U.S. Forest Service, September 18, 1973.

U.S. Department of Agriculture, 1971, *Research and Development Gypsy Moth Program,* Washington, D.C. 22 p.

U.S. Department of Interior, 1970, *The National Atlas of the United States of America,* Washington, D.C. 417 pp.

U.S. Forest Service, 1958, Timber resources for America's future, Forest Resource Report No. 14. Washington, D.C. 713 pp.

U.S. Forest Service, 1970, Evaluating southern pine beetle infestations, State and Private Forestry—Southeastern Area, Division of Forest Pest Control, Atlanta, Ga. 35 pp.

U.S. Forest Service, 1973, The outlook for timber in the United States, Forest Resource Report No. 20, Washington, D.C. 367 pp.

U.S. Forest Service, 1974, Division of Forest Pest Control, Washington, D.C. Personal communication.

U.S. Forest Service, 1975, Pest Management Field Office, Pineville, La. Personal communication.

Veal, M. A., Blair, R. L., Jett, J. B., McKean, W. T., and Cowling, E. B., 1974, Impact of fusiform rust on pulping properties of young loblolly pine, Abstract 185, *in:* Proceedings of the 1974 Annual Meeting of the American Phytopathology Society, Vancouver, B.C.

Waters, W. E., 1969, The life table approach to analysis of insect impact, *J. Forestry* **67**:300–304.

Waters, W. E., 1972, Ecological management of forest insect populations, Proceedings of the Tall Timbers Conference on Ecology of Animal Control by Habitat Management, No. 3, pp. 141–153.

Waters, W. E. and Ewing, B., 1975, Development and role of predictive modeling in pest-management systems—Insects, *in: Longterm and Shorterm Prediction Models of Insects, Phytopathogens and Weed Populations As They Relate to Crop Loss,* Michigan State University (In press).

Wells, D. O., 1969, Results of the southwide pine seed source study through 1968–69, *in:* Proceedings of the Tenth Southern Conference on Forest Tree Improvement, Forest Service, State and Private Forestry, Atlanta, Ga., pp. 117–129.

Wickman, B. E., Trostle, G. C., and Buffam, P. E., 1971, Douglas-fir tussock moth, U.S. Department of Agriculture, Forest Service, Forest Pest Leaflet 86, Washington, D.C. 6 pp.

XI

Progress, Problems, and Prospects for Integrated Pest Management

J. Lawrence Apple and Ray F. Smith

The inequality of food demand and food supply has persisted in parts of the world since the dawn of man's history, but in modern times the populations of developed countries have felt secure in their escape from hunger. This situation changed in 1974 with some food commodities in short supply on a worldwide basis. A high world population growth rate (currently about 2%) and major regional crop failures because of adverse climate and damaging pest attacks (principally insects and diseases) has brought the world feed and food grain reserves to their lowest levels in two decades (Revelle, 1974). Although the actual magnitude of the world food problem is not known, famine is reported in many developing countries, and the death rate is actually rising in at least 12 and possibly 20 such nations in Africa and Southern Asia (NAS, 1975). This imbalance in the world food–people equation has focused unprecedented attention to increased agricultural production in both developed and developing nations. Regardless of whether we succeed or fail in reducing significantly human population growth, the immediate challenge of the United States and the World is to optimize agricultural and other renewable resource productivity per unit of land area, water, fertilizer, energy, and time (Wittwer, 1975). These efforts to increase productivity of the land in both developed and developing countries will accelerate the development and adoption of production practices that gen-

---

J. LAWRENCE APPLE · Departments of Plant Pathology and Genetics, North Carolina State University at Raleigh. RAY F. SMITH · Department of Entomological Sciences, University of California at Berkeley.

erally intensify crop protection problems. The magnitude of agricultural crop losses to pests has not been measured adequately even in the most highly developed countries, but these losses are recognized as being substantial.

In spite of improvements in crop protection technology, there is evidence that the losses due to insect and disease pests have increased in the United States since the 1940s, both in gross amount and as a percentage of crop value (CEQ, 1972). It is thus assumed that improved crop protection has been offset by other changes in crop production technology that were prompted by increased production goals but which have enhanced vulnerability of the crop to pest damage through physical, biological, and genetic changes in the agroecosystem. Factors that have enhanced agroecosystem vulnerability to pests are: (1) greater reliance on ''monocultures'' of major agricultural crops; (2) greater use of fertilizers; (3) improved water management to increase crop yields; (4) multiple cropping; (5) strong selection pressures on yield in developing new varieties which has narrowed the genetic base of principal crops; (6) reduced tillage; (7) introduction of disease, insect, and weed pests into new areas; and (8) dependence on chemical pesticides and other single tactic approaches to crop protection.

The pre- and postharvest crop losses due to pest damage in the developing countries probably average 30% of potential production, but often are much higher on some crops in certain areas. As an illustration, data compiled by the Peruvian Ministry of Agriculture for 1970 (Apple and Smith, 1973) indicated that the loss in potential production due to all classes of pests was 38%, with the breakdown as follows: diseases—11%; insects—16%; nematodes—6%; and weeds—5%.

Achievement of higher levels of adoption for modern agricultural production practices in the developed nations and especially the introduction and adoption of ''green revolution'' technology in the developing nations will bring increasing pressure on crop protection programs. There is already evidence that the transition of ''traditional agriculture'' to modern methods in the developing nations is accompanied by an intensification of pest problems (Apple, 1972; Smith, 1972). Since modern agricultural practices require higher capital investments and since the world food balance is critical, the fluctuation in output that may be caused by massive pest attacks cannot be tolerated. Consequently, more comprehensive, effective, and efficient systems of crop protection must be developed and implemented on a worldwide basis.

It is from this milieu of increased cognizance of world food problems, of increasing losses in potential food production due to pests, of enhanced knowledge concerning the ecology and biology of individual pests, and of the inadequacies and hazards of present approaches to and delivery systems for crop protection that the motivations for new crop protection tactics and strategies have emerged. A small but scientifically strong international group of scientists have for several years provided philosophical and programming leadership for these new tactics and strategies under the rubric of integrated pest management (IPM). They have at times been accused of "bandwagon" tactics and of applying new jargon to long-established practices, but they have demonstrated that the application of ecological principles to crop protection in the form of IPM is sound strategy. As a result, IPM is now finding a place in agricultural production systems throughout the world.

Huffaker and Croft (1975) have described a series of phases in the evolution of an IPM program. These phases reflect the fact that IPM programs generally have a simple beginning. The first approach is an approximation of an ideal system that is subjected to field testing. The difficulties encountered identify problems that must be researched, and in that way IPM systems can be developed step by step. A brief description of the phases of Huffaker and Croft will be useful in subsequent discussions of specific programs or activities.

1. *Single-tactic phase*. Emphasis is generally placed on control of a single pest utilizing a single tactic. This phase does not represent IPM, but the limitations in this approach may lead to its development.

2. *Multiple-tactics phase*. This phase embraces a variety of tactics—cultural, mechanical, physical, chemical, biological, host resistance, regulatory, etc.—in manipulating pest populations. Many IPM systems have not gone beyond this phase.

3. *Biological monitoring phase*. This phase introduces monitoring of pest, natural enemy, and host plant (phenology) populations as the basis for timing the application of various control tactics. This procedure was initiated with the early cotton scouting of 50 years ago but has become greatly elaborated and sophisticated in recent years, as exemplified by the extension pilot pest management projects discussed in this chapter.

4. *Modeling phase*. This involves the conceptualization of the processes involved in pest management systems through mental, pictorial, flowchart, and mathematical models. The earliest integrated control programs depended solely on mental models, but as the volume and complexity of data increased, more sophisticated modeling techniques became necessary.

5. *Management or optimization phase*. This process involves the construction of a functional IPM system utilizing compatible subsystems in optimizing the integration of this IPM system with the overall crop production system. In the past this has been done intuitively, but now mathematical IPM models utilizing techniques for economic analysis and optimization can be utilized.

6. *Systems implementation phase*. This is the ultimate phase through which optimal systems are unified for delivery to and utilization by the grower. According to Huffaker and Croft, this ultimate phase of IPM system development has not been achieved.

Implementation Status of Integrated Pest Management in the United States

Other chapters in this volume refer to various IPM research and extension projects in the United States and elsewhere. We will make brief reference to some of the significant IPM activities in the United States other than those treated in other chapters and comment on the impact on crop protection programs.

A Multi-Institutional IPM Research Project Entitled "The Principles, Strategies, and Tactics of Pest Population Regulation and Control in Major Crop Ecosystems" *

This is one of the most ambitious cooperative research projects ever undertaken in the history of United States agriculture. It involves 18 state universities, the Agricultural Research Service and the Forest Service of the USDA, and industry. The project, popularly known as the "Huffaker Project," is coordinated through the University of California with Dr. C. B. Huffaker as Project Director and Dr. R. F. Smith as Associate Project Director. It was funded at $1.9 million for the first year (1971–72) through the National Science Foundation (NSF), the Environmental Protection Agency (EPA), and the USDA. It has been funded in subsequent years by NSF and EPA at approximately $1.6 million annually. These grant funds, however, represent only a portion of the total cost of this project since each participating institution is providing substantial support from other funding sources. The project will continue in its present form through February, 1977. Although commonly referenced as an "integrated pest management" project, it focuses principally on insect pests of six major crop or forest ecosystems in the United States, i.e., cotton, citrus, alfalfa, pome and stone fruits, soybeans, and pine (bark beetle). Mathematical modeling and systems analysis have received major attention under the project and have provided the research planning matrix through which critical ecosystem relationships and interactions are discovered, data voids identified, and research priorities established.

A review of project results after nearly four years of activity reveals significant accomplishments. The project has demonstrated that major research projects comprising multidisciplinary teams can be implemented and carried out successfully on an interinstitutional and interagency basis. Acceptance and en-

* This information has been obtained from the "Project Reports"; Smith *et al.*, 1974; Smith and Huffaker, 1973; and through the personal involvement of the authors.

thusiastic pursuit by participating scientists attest to its validity. Replication of this research format in other projects, both state and federal, further attests to the distinction gained by this project through its research output. Significant progress has been made in applying advanced technology and systems science to the development of descriptive and predictive models as the basis for deploying pest control tactics, including a more rational use of pesticides. Management systems developed for the major insect pests under study show great promise in pilot IPM extension programs now underway in several states, and some of them have demonstrated a substantial reduction of pesticide usage as compared to presently used traditional systems. This has been accomplished by selecting management tactics from a variety of ecologically compatible and potentially effective alternatives that include: (1) the establishment of an economic need as the basis for remedial action; (2) plant breeding to develop crop varieties resistant to insects and certain plant pathogens; (3) use of parasites, predators, and pathogens (conventional biological control); (4) more selective use of conventional insecticides and fungicides to achieve compatibility with modified programs for insect IPM; (5) new types of cultural controls; (6) use of behavior-modifying chemicals (pheromones) both to monitor and reduce pest populations; and (7) improvements in techniques for measuring population dynamics as the basis for predictive modeling and IPM decision-making.

Although not structured primarily as a graduate education activity, this project has provided 40–50 graduate student assistantships. These students have contributed greatly to the research objectives of the project, but of even greater importance they have been exposed to modern crop protection technology and to the philosophy of an ecological approach to the management of pest populations. This experience has molded their professional attitudes and philosophies, and they undoubtedly are destined to have a major impact on the crop protection sciences in the future.

Since the research base for IPM, especially the modeling component, is necessarily long-term, the principal benefits from modeling and systems analysis are probably yet to come.

Pilot Pest Management Research Program of the Agricultural Research Service (ARS)–USDA

This program was funded during 1971–72 by the USDA for development of new tactics of pest suppression and detection through large-scale trials. It was implemented largely as a series of "in-house" projects within ARS, but some of the activity involved research through cooperative agreements with state agricultural experiment stations. In 1974, 25 projects were in process, and the program was funded at $1.6 million annually. This activity has provided a number of important accomplishments as illustrated by the following (Klassan, 1975): (1) advanced the use of gypsy moth (*Porthetria dispar*) pheromone for detection and population suppression, (2) improved the technology for combating the boll weevil (*Anthonomus grandis*) of cotton through ecological studies, and (3) demonstrated that a plant growth regulator is effective in reducing overwintering of the pink bollworm (*Pectinophora gossypiella*) on cotton. These and numerous other pest management tactics have emerged from this pilot research program. Further, these research efforts helped demonstrate, within federal agencies and state institutions, the validity of the IPM approach to crop protection.

Pilot Pest Management Implementation Projects (Extension)

An important incentive for the development and implementation of IPM programs in the United States was the initiation in 1971 of the pilot pest management implementation projects by two federal agencies, the Animal Plant Health and Inspection Service and the Federal Extension Service, and the agricultural extension services of the land-grant universities. All of these project activities were carried out cooperatively with the states and were funded jointly by federal grants, state funds, and participating farmers. These pilot projects initially involved only insect pests, but some of them have evolved to include all classes of pests consistent with the broadened IPM concept. During 1974, there were 39 pilot pest management projects in 29 states concerned with the following crops and pests (Good, 1975): insects of cotton—14; insects and weeds affecting corn—6; insects and weeds affecting grain sorghum—4; insects, diseases, and weeds affecting peanuts—2; insects and diseases of fruit—6; insects of vegetables and potatoes—4; insects of alfalfa—2; and insects, diseases, weeds, and suckers of tobacco—1.

The pilot projects have introduced large numbers of research and extension crop protection specialists to IPM principles and have stimulated a broad base of interest in and support for IPM programs. As initiated and in their current status of implementation, the above projects represent a wide range of sophistication and application of IPM principles, which is a reflection of the complexity of the pest systems involved and the status of the research data base applicable to their control. Considerable research has been conducted in the past on each of the target pests of the pilot projects, but this has been largely disciplinary research without effective agroecosystem integration. This has left data voids that have in some instances deterred the effectiveness of the pilot projects.

Most of the pilot projects have been limited to one or two classes of pests;

in fact, all of them have involved one or more insect pests but few were directed at all classes of pests. We recognize that implementation of a full-scale IPM program might best be phased over time with initial attention to one or two economically important pests of a single class (e.g., insects) and then evolving to include other classes (e.g., weeds and pathogens). This has been the history of several of the pilot projects. But many of these projects are still too restricted in scope and lack the interdisciplinary collaboration and integration required to achieve the optimal condition. Some of them have not evolved beyond the ''single-tactic phase'' of Huffaker and Croft (1975).

Good (1975) summarized the experiences of the 39 pilot pest management projects as follows: (1) growers are now recognizing the advantages of IPM; (2) most growers will apply pesticides on an as-needed and timely basis when properly advised; (3) practices that have indirect or delayed effects and which represent additional production costs, such as diapause control or crop destruction, are not generally adopted on a free-choice basis but require regulatory authority; (4) growers have become more aware of the economics of crop protection practices and of the need to utilize economic guidelines in making IPM decisions; (5) progressive growers are willing to pay for improved IPM advisory services by participating in grower cooperatives or by contracting a private IPM consultant; and (6) in order to be both practical and economical, an IPM program should include all the important crops in a given area that have common pests and should include all the important pest classes—weeds, insects, diseases, and nematodes.

IPM Programs Through the State Agricultural Experiment Stations

A principal source of research results that are critical to the development of IPM programs has been the state agricultural experiment stations, a system established through the Hatch Act of 1887 and partially funded by federal grants appropriated under authority of that act. Historically, much of the crop protection research has been disciplinary-oriented, but without this knowledge base on individual pests, the development of IPM programs would not be an option. The multidisciplinary, ecosystem approach ascribed to IPM does not obviate a thorough knowledge of the biological and physical parameters of the agroecosystem, quite the contrary. The availability of information derived through disciplinary research now makes possible the construction of IPM programs, and the continuation of that research will be essential to the maintenance and further development of IPM.

Programs at several of the state agricultural experiment stations have evolved to include both disciplinary and multidisciplinary research on crop protection problems. As an example, the majority of the ''principle investigator'' scientists contributing to the Huffaker Project are faculty at land-grant institutions, and major costs of the Huffaker Project are provided through the state agricultural experiment stations.

IPM research has also been stimulated through the state agricultural experiment stations by designated federal grants administered through the Cooperative State Research Service of the USDA. The extension pilot IPM projects in the various states have also created an awareness of the need for integrative IPM research to fill the backstopping role required in the respective states.

IPM research in state agricultural experiment stations has now evolved to the point that several research projects have been developed and funded with existing resources. There are also several ''regional'' research projects within the experiment station system, wherein two or more states have combined their resources to research a common problem. In response to motivation from crop protection scientists in the land-grant institutions and from grower groups, IPM research has become ''institutionalized'' in the agricultural experiment station system and undoubtedly will persist and flourish as a recognized research problem area.

Pest Management Curricula in the Land-Grant Universities

The IPM philosophy has had major impact on instructional programs in crop protection at the land-grant universities in the United States. At least 22 of these institutions now have undergraduate curricula either in plant protection or pest management, 3 have M.S. programs, and several others have undergraduate and graduate (both M.S. and Ph.D.) programs under consideration.* The land-grant institutions anticipated a demand for persons broadly educated and trained in the ecological approach to crop protection and are responding with the creation of new educational programs.

Many of the undergraduate academic programs in plant protection or pest management were initiated cooperatively by combinations of departments representing plant pathology, entomology, and weed science. The first multidisciplinary plant protection curriculum was initiated at North Carolina State University in 1959 and included pathogens, insects, and weeds. Most of the curricula established before 1972 were entitled ''plant protection'' and none were entitled ''pest management,'' and most of them comprised existing rather than new courses reflecting the integrated approach to crop protection. Sub-

* Results from a survey conducted by the senior author in October, 1974.

sequent to the 1972 RICOP * sponsored workshop on ''Systems of Pest Management and Plant Protection'' (Browning, 1972), many new undergraduate programs were established. Some of these were entitled ''pest management'' and some of the earlier established curricula were retitled to reflect pest management. The RICOP workshop also stimulated broader disciplinary integration in pest management curricula and contributed to the development of team-taught interdisciplinary courses at several institutions.

The pest management and/or plant protection undergraduate curricula in most of the offering institutions have not attracted large numbers of students. This probably reflects inadequate employment opportunities in broad pest management programs and the potential employment hazard, from the view of the student, of receiving generalist *versus* specialist training and thus not meeting the employment criteria for more specialized positions. However, as action pest management programs are expanded and further developed, the demand for supportive field and laboratory technicians should increase to the advantage of the B.S. graduates in pest management. There are now a number of indications that these employment opportunities are materializing.

Land-grant institutions have been cautious in developing doctoral level programs in pest management. The feasibility has been discussed extensively but without consensus conclusion. There is a growing number of private pest management consultants, and some of them (Cox, 1971) are strong advocates of the need for broad doctoral degree training in pest management (perhaps a professional rather than a Ph.D. degree). But others feel that the traditional Ph.D. programs in the crop protection disciplines offer sufficient flexibility to accommodate a good educational base in pest management. As pest management action programs evolve and as the role of the private sector in the pest management delivery systems become more clearly defined, there will undoubtedly be increased opportunities for doctoral level pest management specialists; however, we believe that the traditional Ph.D. degree programs of the land-grant universities in crop protection disciplines will require considerable modification to provide the training experience appropriate to this need.

Implementation Status of Integrated Pest Management Programs Outside the United States

IPM activities are not unique to the United States. Operational programs have been developed in several of the agriculturally advanced countries and in some of the developing countries. We shall make no effort to present a compre-

* RICOP is the Resident Instruction Committee on Policy of the Division of Agriculture, National Association of State Universities and Land-Grant Colleges.

hensive inventory of such programs but refer to some of the more significant international programs.

Programs of the International Organization for Biological Control in Western Europe

For about 15 years, the West Palaearctic Regional Section of the International Organization for Biological Control (IOBC) has been taking the leadership in Western Europe in the development of IPM (Braber, 1974 and 1975). Following the early successes obtained in apple orchards, pest situations in other crops have been studied for the possibility of applying IPM techniques.

The first big step forward in apple orchards was the establishment of economic thresholds of the major pests through close international collaboration within the framework of IOBC. Based on these, programs were set up where control measures were taken only after assessment of the pest population. These supervised control programs allowed a reduction of 50% of the quantity of pesticides normally used while the quantity and quality of the crop was maintained. Later, selective methods for control of specific pests were introduced.

The IOBC Working Group in IPM distinguished three phases of development. The first research phase involved only research workers and consisted of an analysis of the pest situation and research on the possibilities of applying different control techniques and the construction of an IPM model. The second development phase involved testing of the model under field conditions to determine limitation and economic feasibility. During this phase, establishment of collaboration between research and extension service workers was essential. The third or application phase involved providing direct assistance to farmers. The main responsibility during the application phase must be taken over by an extension service agency, although a permanent dialogue between research workers and extension technicians will be needed. In addition to apple orchards, the application phase has been reached in the IOBC program with peaches, citrus, cereal, sugar beets, and glasshouse crops. Surely, the IPM approach of IOBC has had and continues to have a very significant impact on evolution of crop protection in Europe.

Programs of the Food and Agriculture Organization (FAO) of the United Nations

FAO initiated IPM * activities in 1963 in response to the growing concern about the undesirable side-effects of large-scale organic pesticide usage and the

* FAO used "integrated pest control" terminology to reference IPM-type activities. For purposes of this discussion, IPM and integrated pest control will be considered synonymous.

failure of this approach to provide economic pest control. An early significant activity was the organization of a symposium in October, 1965 (FAO, 1966) that stressed the urgency of promoting and developing IPM and which recommended the creation of a body of experts in this field. The Panel of Experts on Integrated Pest Control was established by the Director-General of FAO in 1966. The purposes of the FAO Panel are to advise the Director-General on matters pertaining to IPM policy and programs within FAO; to maintain purview of the evolution of IPM (principles, procedures, and techniques); to promote joint IPM research on major pests of international importance; and to collate, interpret, and disseminate information on IPM research and development. The Panel comprises a membership of 38 scientists appointed in their personal capacities for four-year terms. During the course of the six Panel sessions that have been held to date, IPM has been defined and general procedures for establishment have been suggested, the status of IPM research and implementation throughout the world has been reviewed, and recommendations for various actions have been made to FAO and its member governments.

FAO has developed and implemented 13 field projects on high-priority IPM problem situations in developing countries with the financial assistance of the United Nations Development Program (UNDP) and other organizations. The first of these projects was implemented in 1964. More than 80 international crop protection specialists have been involved in these projects.

At its Fourth Session (December, 1972), the Panel of Experts on Integrated Pest Control proposed a global project on integrated pest control as a followup to Recommendation No. 21 of the 1972 United Nations Conference on the Human Environment in Stockholm (FAO, 1973). A proposal for the establishment of a global project was submitted by FAO to the United Nations Environment Program (UNEP) in 1973. UNEP endorsed the concept and agreed to support preliminary activities. The Panel subsequently developed the framework for a "Cooperative Global Program for the Development and Application of Integrated Pest Control in Agriculture" at an *ad hoc* Session in October, 1974 (FAO, 1975). The Global Program "will aim at promoting the development and application of safer, more effective, and more permanent plant protection procedures and techniques, through the combined use of all compatible methods . . . and will contribute to the development of sufficient expertise in the application of the integrated pest control concept, so that developing countries will be able to develop and carry out integrated control programs for pests of major economic importance. Long-standing support by FAO for such programs will be expanded by direct assistance to national research and advisory bodies and, in particular, by continuing to help coordinate regional programs." The following criteria were developed for identifying priority crops under the Global Program (FAO, 1975): (1) vital national and regional importance to the economy and well-being of human populations, (2) seriousness of crop losses caused by pests and diseases, (3) environmental problems and impairment of pest control created by undue reliance on pesticides, (4) impending problems from rapidly accelerating pesticide usage in response to urgent needs to increase food production, and (5) realized or potential success in integrated control practices. In applying these criteria, the Panel gave highest priority to cotton, rice, and maize/sorghum. Other crops (cole, potatoes, grain legumes, cassava, sugar cane, and coconut palms) were also selected for inclusion in the program at a later stage. The Coordinator for the Global Program has been employed by FAO, and implementation funds are now being sought.

The extensive nature and visibility of the complex of FAO-sponsored programs in IPM and related activities have given great impetus to the concept on an international basis.

The Pest Management and Related Environmental Protection Project—University of California

This is a contract project (briefed as the UC/AID Project) funded by the United States Agency for International Development (USAID) through the University of California at Berkeley with Dr. Ray F. Smith as Project Director. It was initiated in 1971. Other U.S. institutions or agencies participating in the Project through consultants and subcontracts have been Cornell University, University of Miami School of Medicine, University of Florida, North Carolina State University, and the U.S. Department of Agriculture.

The principal Project goals are to provide developing countries with assistance in devising and implementing ecologically sound and economically valid IPM systems for the economic control of agricultural pests (insects, pathogens, and weeds) so that their long-term agricultural productivity goals can be achieved. These goals are to be attained by developing their in-country scientific and institutional capacity to handle diverse pest problems in the following manner: (1) through training and retraining of crop protection personnel from participating countries; (2) establishment of technical assistance and extension projects aimed at specific crop protection problems; and (3) assisting local personnel and their institutions in establishing or improving research, training, and extension programs in crop protection. The Project is complementary to the commodity improvement projects funded by USAID, the international agricultural research network funded through the Consultative Group, and food crop improvement projects financed by other donors for purposes of increasing the food supply in the developing nations. (The latter objective cannot be realized until the major pests of important food crops are identified, studied, and brought under management control.)

The UC/AID Project activities have included the following: (1) provided advisory services to assist AID/Washington in developing improved pesticide evaluation, procurement, and use protocols; (2) served as a backstopping resource for USAID country missions in the area of pest management; (3) developed and implemented on a permanent basis a procedure for backstopping and providing research and technical assistance to USAID missions in the evaluation, procurement, and use of pesticides; (4) assisted countries in developing safeguards and regulatory procedures for the importation, manufacture, formulation, distribution, and use of pesticides; (5) collaborated with countries in developing national regulatory and pesticide use monitoring systems; (6) identified high-priority regional pest problems on the basis of their damage potential to food crops; (7) assisted countries in developing research and training procedures for development of scientific and technical skills; (8) collaborated in planning country-based IPM and environmental protection systems; and (9) interfaced country-based systems with the international cooperative research and technical assistance network.

It should be noted that activity (6) (determination of high-priority regional pest problems) was carried out by six multidisciplinary study teams that made visits to thirty-two developing countries during 1971–1972. Each team comprised four or five experienced scientists representing the disciplines of plant pathology, entomology, weed science, and nematology. Each survey team prepared a report of its findings that included: (1) an analysis and priority ordering of the pest problems encountered; (2) recommendations for research, training, and control programs based on the importance of the problems and the probability of successful control; (3) evaluation of the technical capability of local educational institutions and agencies; (4) identification of individuals who could qualify for project training programs or as candidates for degree training; and (5) evaluation of the social and cultural problems that might deter the development and implementation of appropriate IPM programs. Subsequently, the UC/AID Project sought funds to implement projects around some of the high-priority problem areas identified by the study teams. One of these—a worldwide project on the root knot nematode (*Meloidogyne* sp.)—has now been funded, and others are under consideration by AID/Washington. These study team activities have made a very positive contribution to the furtherance of IMP program development both in developing countries and in the United States. Firstly, they brought together a large interdisciplinary group of senior crop protection scientists from U.S. institutions. Their interaction as team members will have a lasting catalytic effect on IPM program development in some of their own institutions. Secondly, the study team reports represent perhaps the most comprehensive single effort to identify and document the high-hazard crop protection problems in the developing countries. This has reinforced the recognition of these problems by local authorities and their determination to give adequate attention to them.

The survey teams noted frequently the mismanagement of pesticides in the developing countries. This fact along with direct requests from several developing countries led to the organization of Pesticide Management Seminars or workshops in cooperation with local agencies and the World Health Organization which have been conducted in El Salvador, Indonesia, and the Philippines. Host country participation has come from the ministries of agriculture and public health. Other workshops are planned. Follow-up activities to those already conducted include: (1) the organization of agromedical teams in each country with purview over pesticide importation, registration, use, and residue monitoring programs; (2) assistance with development of reliable pesticide residue analytical laboratories; (3) development of protocols for managing pesticides from time of manufacture or importation through ultimate use or disposal; and (4) development of chemical impregnation methods for clothing to protect agricultural workers against pesticide poisoning. These workshops have done much to enhance the understanding of pesticide use problems in the participating countries and to encourage appropriate action at the highest levels of authority. They should ultimately have a desirable impact on both food safety and environmental quality in the developing countries. The actions being pursued through these workshops are necessary components of any effective IPM system for agricultural crops.

Problems and Prospects for Developing Integrated Pest Management Programs

The programs and activities described in other chapters in this volume as well as those briefed above would indicate that IPM, as a relatively new organized activity within the agricultural sciences, is vigorous and moving strongly toward becoming a generally accepted concept and permanently established as the "capstone" crop protection discipline. But there are deterrents to the further development and adoption of IPM, both in the United States and elsewhere.

A principal deterrent is the fact that IPM is multidisciplinary, and its pursuit requires that several professional scientists collaborate in program planning and execution. Some scientists prefer to work independently and choose not to subject themselves to the compromising experiences often inherent in collaborative programs. Further, the language (and possibly the concepts although there is argument on the point) of IPM are unfamiliar to many well established crop protection scientists and thus they choose not to tread in unfamiliar territory.

IPM was first articulated within the ranks of entomology and gained considerable attention and funding as an insect management approach before the concept was expanded to include all classes of pests impinging on the crop ecosystem. This fact of origin has tended to alienate some plant pathologists, for example, because certain descriptive details and management tactics accorded the concept by entomologists are not directly applicable to plant pathogens. Further, some contend that the term ''pest'' has become synonymous with ''insect'' through precedent of usage. These factors have prompted the attitude by some that plant pathologists and plant pathology will lose identity if amalgamated under the rubric of IPM.

Misunderstandings of the IPM concept and its underlying principles is a major deterrent to its adoption and application. Chiarappa (1974) assessed the possibility of ''supervised plant disease control'' (SPDC) in pest management systems and concluded that ''there appear to be only a rather limited number of diseases which are actually suitable to SPDC.'' Chiarappa developed criteria that should be satisfied by a disease before SPDC would be justified. Before a disease problem would be suitable for SPDC, a chemical control method must be available, a high-value crop must be involved, the cost of chemical control must be high, the cost of monitoring must be low, the ''incubation period'' between the time of infection and resulting crop damage must be sufficient to permit the disease to be arrested, presence of the disease must be evidenced in above-ground plant parts, the risk hazard of SPDC to the grower must be low, and the grower must have the capability for timely treatment of necessary acreage. In applying these criteria to the stem rust disease (*Puccinia graminis tritici*) of wheat, as an example, Chiarappa concluded that this disease was not suited to SPDC because it is a low-value crop per unit of land, and he pointed out that stem rust is usually controlled by use of disease-resistant varieties. An obvious limitation of Chiarappa's concept is that it assumes chemical control as the only management tool. If this limitation is removed, we must conclude that the stem rust disease of wheat in the United States has been under a very sensitive and intensive management system for many years (Rowell and Roelfs, 1974). The pathogen spore populations are monitored continuously each year as to race composition, and these data are used in deploying the resistant germ plasm of the host plant. Perhaps stem rust of wheat is one of the best examples of ''disease management'' in plant pathology. It represents a regional approach to the management of a regional problem which is a fundamental tenet of the IPM concept. We must conclude that Chiarappa's restrictive SPDC concept is inconsistent with the IPM concept and that he has understated the applicability of the latter to the management of plant pathogens.

Another point of controversy is the notion that IPM was conceived as an alternative approach to the use of chemicals and that it views pesticides as a negative factor in agricultural production. Some in the pesticide industry view this notion both as a threat to the agricultural production systems of advanced nations and to the marketing of chemical pesticides (Grimes, 1975). If viewed as a system to reduce substantially the use of pesticides in agriculture, undoubtedly the chemical industry would seek to deter the wide-scale implementation of IPM. But we counter this assumption, firstly, with the firm assertion that IPM is not antipesticidal in mission. It may result in reduced pesticide usage, but only if these materials are now being used in excess or as substitutes for other more effective and economical methods. Secondly, if IPM does hold potential for improved pesticide usage, all segments of the agricultural enterprise and the consumer will benefit in the long run.

The misconception of IPM as a panacea applicable to all pest problems also stirs conflict. IPM is not a panacea; further, it is not a fixed, patterned approach but a concept sensitive to the needs of the problem situation. There are probably only a few crop ecosystems and pest complexes that can justify or would require application of the total spectrum of pest management strategies and tactics. The management strategy must be developed for each pest complex within each crop ecosystem, and that strategy may be conditioned year to year by prices of commodities and purchased inputs, legislation, social issues, and other factors. Many crop production enterprises cannot justify the resource cost requisite to a sophisticated IPM system. But regardless of crop enterprise scale or nature of the pest problem, the ecological principles implicit in IPM can be applied to the limits of economic justification.

The mislabeling of some traditional crop protection projects as IPM has done much harm to the image and has diluted the performance record of valid IPM projects in the eyes of some decision-making persons in the United States. Some of the funds appropriated for action IPM programs through the extension services have been used to support project activities that could hardly qualify as IPM. Unfortunately, comparable and parallel funding was not allocated to the counterpart agricultural experiment stations in the "pilot project" states; consequently, the research backstopping component was often deficient. In many situations, these extension projects did not evolve to higher levels of IPM achievement because the critical companion research was not funded. These facts have given the critics of IPM a substantive basis to oppose continuation of federal appropriations for support of both extension and research IPM projects.

Application of the IPM concept is often based on an estimated population density and on the probability of resultant economic damage. Since these probabilities are based not only on the pest population density but also on other biological and physical factors of the agroecosystem, the accuracy of the projected loss will not be 100% correct. Although many benefits will accrue to farmers and society at large from the application of IPM, individual farmers could incur economic loss resulting from an erroneous pest population or loss estimate. In order to relieve the individual farmer of this loss hazard, some type of "crop protection insurance" must be available to protect against economic loss. This is a dimension of IPM that has not been provided by insurance underwriters except in a few areas. Such protection must be available at a cost, when added to that of the IPM program, that is competitive with the conventional chemical control approach. Lack of such insurance could deter the adoption of IPM by many farmers who are not willing to hazard the risk of action decisions based upon probabilities.

We have pointed out these several factors above that have deterred the implementation of IPM research and extension programs in the United States. We are convinced, however, that the most important of these deterrents are political and parochial factors rather than faults of the IPM concept. We are encouraged by the implementation record for IPM to date and predict that it will become increasingly important as an agroecosystem management strategy.

Literature Cited

Apple, J.L., 1972, Intensified pest management needs of developing nations, *BioScience* **22:**461–464.

Apple, J. L., and R. F. Smith, 1973, Crop protection problems in Latin America, *Devl. Dig.* **10(4):**98–105.

Brader, L., 1974, Present Status of Integrated Control of Pests, *Mededeelingen Fakulteit Landbouw, Wetenschappen Ent.* **39(2):**345–365.

Brader, L., 1975, Progress in Biological and Integrated Control, *Bull IOBC/WPRS* **1975/1.** 152 pp.

Browning, C. B., 1972, Systems of pest management and plant protection, Report of the RICOP Committee on Plant Protection, St. Louis, Mo. 24 pp.

Chiarappa, L., 1974, Possibility of supervised plant disease control in pest management systems, *FAO Plant Prot. Bull.* **22:**65–68.

Council on Environmental Quality (CEQ), 1972, *Integrated Pest Management,* US. Government Printing Office, Washington, D.C. 41 pp.

Cox, R. S., 1971, *The Private Practioner in Agriculture,* Solo Publications, Lake Worth, Fla. 192 pp.

Food and Agriculture Organization (FAO), 1966, Proceedings of FAO Symposium on Integrated Pest Control, Rome, October, 1965, Vols. 1–3.

Food and Agriculture Organization (FAO), 1973, Report of the Fourth Session of the FAO Panel of Experts on Integrated Pest Control, Rome, December 6–9, 1972. 35 pp.

Food and Agricultural Organization (FAO), 1975, The development and application of integrated pest control in agriculture, *in:* Report on an *ad hoc* Session of the FAO Panel of Experts on Integrated Pest Control, Rome, October 15–25, 1974. 24 pp.

Good, J. M., 1975, Integrated pest management projects: Progress and future programs, *in:* Proceedings of Peanut, Tobacco and Vegetable Pest Management Workshop, Raleigh, N.C., February 18–20, 1975, pp. 123–129.

Grimes, W., 1975, The role of the entomologist in food and fiber production, *J. Ga. Entomol. Soc.* **10(Suppl. I):** 1–7.

Huffaker, C. B., and Croft, B. A., 1975, Integrated pest management in the USA—Progress and promise, *Environ. Health Persps.* (In press).

Klassan, W., 1975, Pest management: Organization and resources for implementation, *in: Insects, Science, and Society* (*D. Pimental, ed.*), Academic Press, New York, pp. 227–256.

National Academy of Sciences (NAS), 1975, Population and food—Crucial issues, Committee on World Food, Health and Population, National Academy of Science, Washington, D.C. 50 pp.

Revelle, R., 1974, Food and population, *Sci. Am.* **231(3):**160–170.

Rowell, J. B., and Roelfs, A. P., 1974, Wheat stem rust, *in:* USA-USSR Symposium on Long-Term Prediction Models of Insects, Phytopathogens, and Weed Populations as They Relate to Crop Loss, Michigan State University, October 15–17, 1974. 24 pp.

Smith, R. F., 1972, The impact of the green revolution on plant protection in tropical and subtropical areas, *Bull. Entomol. Soc. Am.* **18:**7–14.

Smith, R. F., and Huffaker, C. B., 1973, Integrated control strategy in the United States and its practical implementation, *OEPP/EPPO Bull.* **3(3):**31–49.

Smith, R. F., Huffaker, C. B., Adkisson, P.L., and Newsom, L. D., 1974, Progress achieved in the implementation of integrated control projects in the USA and tropical countries, *OEPP/EPPO Bull.* **4(3):** 221–239.

Wittwer, S. H., 1975, Food production: Technology and the resource base, *Science* **188:**579–584.

Index